AF292837

Discrete and Fractional Programming Techniques for Location Models

COMBINATORIAL OPTIMIZATION

VOLUME 3

Through monographs and contributed works the objective of the series is to publish state of the art expository research covering all topics in the field of combinatorial optimization. In addition, the series will include books which are suitable for graduate level courses in computer science, engineering, business, applied mathematics, and operations research.

Combinatorial (or discrete) optimization problems arise in various applications, including communications network design, VLSI design, machine vision, airline crew scheduling, corporate planning, computer-aided design and manufacturing, database query design, cellular telephone frequency assignment, constraint directed reasoning, and computational biology. The topics of the books will cover complexity analysis and algorithm design (parallel and serial), computational experiments and applications in science and engineering.

Series Editors:

Ding-Zhu Du, *University of Minnesota*
Panos M. Pardalos, *University of Florida*

Advisory Editorial Board:

Afonso Ferreira, *CNRS-LIP ENS Lyon*
Jun Gu, *University of Calgary*
D. Frank Hsu, *Fordham University*
David S. Johnson, *AT&T Research*
James B. Orlin, *M.I.T.*
Christos H. Papadimitriou, *University of California at Berkeley*
Fred S. Roberts, *Rutgers University*

The titles published in this series are listed at the end of this volume.

Discrete and Fractional Programming Techniques for Location Models

by

Ana Isabel Barros

TNO Physics & Electronics Laboratory,
Division Operational Research & Business Management,
Section Air Defence,
The Hague, The Netherlands

SPRINGER-SCIENCE+BUSINESS MEDIA, B.V.

A C.I.P. Catalogue record for this book is available from the Library of Congress.

ISBN 978-1-4613-6824-3   ISBN 978-1-4615-4072-4 (eBook)
DOI 10.1007/978-1-4615-4072-4

Printed on acid-free paper

To the ones I love ...

Contents

List of Figures

List of Algorithms

List of Tables

Acknowledgments

This book is a revised version of my Ph.D. thesis, [9], that was written at the Econometric Institute, Erasmus University Rotterdam, the Netherlands. The first three years of the Ph.D. research were sponsored by the Portuguese grant committee *Junta Nacional de Investigação Científica e Tecnológica* under contract BD/707/90-RM, while the last year was sponsored by the *Tinbergen Institute of Rotterdam*.

In the initial stage of the Ph.D. research I worked mostly with Martine Labbé who made me realize that location analysis can be a fascinating topic. During the last two years, Hans Frenk was my daily supervisor. His knowledge, encouragement and energy kept me "moving forward". Although I learned a lot with him, Rockafellar is still not my favorite bed time book! Hans also introduced me to Shuzhong Zhang and Siegfried Schaible. Our collaboration had an important impact in my research. Finally, I would like to express my gratitude to my *promotor*, Alexander Rinnooy Kan. Although, extremely busy he always found the time to give me valuable suggestions and guidelines to my research.

After finishing the Ph.D. I went back to the *Departamento de Estatística e Investigação Operacional* of University of Lisbon for one year, after which I returned to the Netherlands. Meanwhile, I was approached by John Martindale with the idea to republish my Ph.D. thesis by Kluwer. I began revising my thesis without suspecting that it would take me quite a while to become satisfied with the final result. I would like to acknowledge John's enormous patience handling the whole process of publishing this work.

Lots of people had decisive influence in the process of writing both the thesis and this book. Unfortunately, if the acknowledgments of the thesis were already rather extensive they only became longer in the last years. Hence, hopping not to upset anybody I am condensing them! I would like to thank the Econometric Institute and in particular the *Mabes* group. Also a special thanks to the Ph.D. students who welcomed me at the Institute, after completing my Ph.D., as one of the "gang"! Moreover, I also want to acknowledge Tareca, Quim and Marcel for their support and technical help.

To my Dutch friends and family, especially *Ma* Blok, my deep gratitude for making Holland a really cozy new home!

When I wrote the thesis I soon realized that I could not find appropriate words to thank my parents and Auke. Three years later I am faced with the same problem. Nevertheless, I would like to express again my gratitude and appreciation to the *Best Parents in the World* and to the two diamonds of my life: Auke (the tallest for the moment) and Artur!

Notation

$\vartheta(P)$	optimal value of problem (P)
$(\overline{P})$	linear relaxation of problem (P)
∞	infinite
$\emptyset$	the empty set (without elements)
$I\!N$	the set of positive or zero integers
$I\!N^n$	Cartesian product of the set $I\!N$ with itself, n times
$I\!R$	the set of real numbers
$\overline{I\!R}$	$I\!R \cup \{-\infty, +\infty\}$
$I\!R^n$	Cartesian product of the set $I\!R$ with itself, n times
$I\!R^n_+$	positive orthant of the space $I\!R^n$
$\mathcal{X} \times \mathcal{Y}$	Cartesian product of sets $\mathcal{X}$ and $\mathcal{Y}$
$\mathrm{co}(\mathcal{X})$	the convex hull of set $\mathcal{X}$
$\mathrm{ri}(\mathcal{X})$	the relative interior of set $\mathcal{X}$
$(x)^+$	corresponds to the maximum between 0 and x
$|x|$	the absolute value of x
∇f	the gradient of function f
∂f	the subgradient set of function f
bold capitals	matrices
$A^\top$	the transpose of matrix A
bold lowers	vectors
x_i	ith component of vector x
$\mathbf{0}$	the null vector
$\|x\|$	the Euclidean norm of the vector x

Σ the set defined by $\{y \in I\!R^n : y \geq 0, \sum_{j \in J} y_j = 1\}$

$\mathcal{O}(f(m, n))$ the complexity order of an algorithm, which applied to a problem whose size is measured by the parameters m, n requires a computing time $T(m, n) \leq k f(m, n)$, where f is a function of m and n and k a positive constant.

$\boxed{One}$

Introduction

At first sight discrete and fractional programming techniques appear to be two completely unrelated fields in operations research. We will show how techniques in both fields can be applied separately and in a combined form to particular models in location analysis.

Location analysis deals with the problem of deciding where to locate facilities, considering the clients to be served, in such a way that a certain criterion is optimized. The term "facilities" immediately suggests factories, warehouses, schools, etc., while the term "clients" refers to depots, retail units, students, etc. Three basic classes can be identified in location analysis: *continuous location, network location* and *discrete location*. The differences between these fields arise from the structure of the set of possible locations for the facilities. Hence, locating facilities in the plane or in another continuous space corresponds to a continuous location model while finding optimal facility locations on the edges or vertices of a network corresponds to a network location model. Finally, if the possible set of locations is a finite set of points we have a discrete location model. Each of these fields has been actively studied, arousing intense discussion on the advantages and disadvantages of each of them. The usual requirement that every point in the plane or on the network must be a candidate location point, is one of the mostly used arguments "against" continuous and network location models. Also, in some practical applications it is easier to adapt or to model problems using a discrete location model. In this book we will mainly consider discrete location models. However, a continuous location model in the plane will also be discussed.

An important aspect in location analysis is the criterion considered in the optimization process. In discrete location analysis the traditional criterion is usually to maximize the total net profit. The total net profit measures the gains of a certain decision, i.e. the difference between the sum of the profits of serving each client via certain facilities and the fixed costs associated with those facilities. Clearly, the optimization of this type of location models, with a linear type of objective function, requires the use of classical integer programming techniques. However, in some economical applications it is also important to consider other nonlinear types of criteria, like maximizing the profitability index. Dealing with discrete location problems with this type of criterion, where a ratio of two functions has to be optimized, requires not only the use of classical integer programming techniques but also of fractional programming techniques.

Fractional programming is a special field of nonlinear programming dealing with optimization problems where the objective function consists of a ratio of given functions. An immediate extension of this class corresponds to *generalized fractional programming* where the goal is to minimize the largest of several ratios of functions. Due to the special characteristics of the ratio function, it is possible to derive special solution techniques to tackle this class of nonlinear problems. Although the scope of application of fractional programming is vast, it has not been actively related to specific operations research problems, and in particular to location analysis. A possible explanation for this phenomenon is given by Schaible and Ibaraki [125]

> "... fractional programming gave itself a somewhat questionable reputation in the operations research community by divorcing itself too much from the applications ..."

On the other hand, Sniedovich [129] believes that this reluctance of the operations research community

> "... lies in fact in the mathematical format given to fractional programming in the literature and the thinking embedded therein ... the mathematical phrasing of fractional programming in the literature betrays a rather lean theoretical foundation, which has the effect of divorcing fractional programming from optimization theory at large, thereby

rendering it relatively unknown in operations research ..."

The above explanations stress not only the importance of considering applications of fractional programming within operations research, but also of providing a more solid interpretation of known results in this field. These remarks led to the analysis of applications of fractional programming to discrete location and of generalized fractional programming to continuous location. We also attempted to provide a clear and geometrical interpretation of the basic techniques of generalized fractional programming. This geometrical interpretation combined with the use of Lagrangian duality enabled the development of new theoretical results and solution techniques for generalized fractional programming.

The structure of this book reflects the above line of though. Hence, Chapter 2 is devoted to discrete location, in particular to the maximization of the total net profit of location problems with two levels. As mentioned before, our interest in fractional programming led to the search of possible applications of this field in discrete location. Therefore, Chapters 3 and 4 are dedicated to fractional and generalized fractional programming. We start by discussing, in Chapter 3, some basic results on classical fractional programs and relate them to discrete fractional location problems. As this chapter shows, it is extremely important to have a sound knowledge of fractional programming when solving a fractional location model. Moreover, the results contained in Chapter 3 also led to the search for possible applications of generalized fractional programming in location analysis. As a result, we encountered an allocation model arising in continuous location that corresponds to a generalized fractional programming problem. This model as well as an extensive discussion about generalized fractional programming and its tools are contained in Chapter 4. Finally, some final remarks and directions for future research are contained in Chapter 5.

Two

Discrete Location Models

Discrete location deals with deciding where to locate facilities within a finite set of sites, taking into account the needs of the clients to be served in such a way that a given criterion is optimized. In this description there are several vague points, like the existence or not of additional constraints on the facilities, the way clients demand is satisfied, and so forth. Hence, by particularizing these points different models arise. For instance, the existence or not of capacities on the supplies creates either capacitated or uncapacitated models, and the number of levels of facilities to be located yields either 1-level or multi-level location problems.

In this chapter we will focus on discrete uncapacitated location models with two levels of facilities. These models are generalizations of the *uncapacitated facility location problem* and hence, we start by reviewing this classical model. We also illustrate how some of the combinatorial and integer programming techniques can be applied to this classical model. This brief survey will be complemented with a short overview over the multi-level location models which are directly related to the new model presented in Section 2.4. Before analyzing in detail this new model, we will first focus on the submodularity property. This property is particularly important since it enables the derivation of both theoretical and practical results which can be embedded in the available solution methods. Hence, it is interesting to investigate if as in the case of the uncapacitated facility location problem some other location models also satisfy this property. Unfortunately, as shown in Section 2.3 the *multi-level uncapacitated facility location problem* does not satisfy the submodularity property.

The main part of this chapter is devoted to the analysis of a new 2-level location model. This general model considers the problem of locating pairs of facilities and depots in order to maximize the sum of the profits of serving each client from exactly one facility and one depot minus the fixed costs of opening those facilities and depots and also the fixed costs of having facilities and depots operating together. Due to the type of fixed costs involved, this general 2-level model also includes as special cases some known models. In order to tackle this problem we propose in Section 2.4 three different formulations and derive lower and upper bounds using Lagrangian relaxation and heuristics. These bounds are embedded in a branch and bound type algorithm. Computational results for the general model and two of its special cases are also reported.

2.1 The Uncapacitated Facility Location Problem

One of the best known discrete location models is the *uncapacitated facility location problem*. This problem consists of choosing where to locate facilities, selected from a discrete set of possible locations, in order to maximize the profit associated with satisfying the demand of a given set of clients. Usually, setting up a facility involves significant costs that are not proportional to the production level of this facility. Clearly, these fixed costs depend also on the location of the facility. Hence, fixed costs of opening a given facility depending on the location are considered. Figure 2.1 illustrates a solution to the problem of locating depots that provide service to a set of clients.

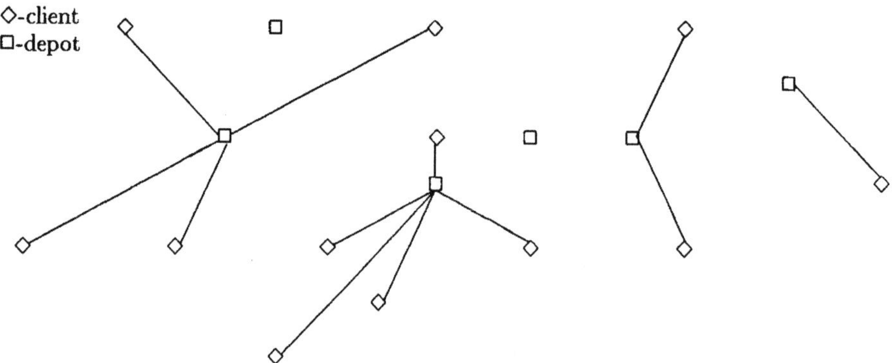

Figure 2.1: An example of a solution to the uncapacitated facility location problem.

Facilities are assumed to have unlimited capacity, i.e. any facility can satisfy the demand of all clients. When each facility can only supply demand up to a given limit, the problem is called the *capacitated facility location problem*.

There exists some controversy over who first formulated the uncapacitated facility location problem, but it is generally accepted that the first formulations of this problem are due to Balinski and Wolfe [8], Manne [89], Kuehn and Hamburger [86], and Stollsteimer [130].

The problem can be formalized as follows. Let $I := \{1,\ldots,m\}$ denote the set of clients and $J := \{1,\ldots,p\}$ the set of sites where facilities can be located. Let also f_j denote the fixed cost of opening facility j, and c_{ij} the profit associated with satisfying the demand of client i from facility j. Usually, c_{ij} is a function of the production costs at facility j, the demand and selling price of client i and the transportation costs between client i and facility j. Without loss of generality, we will assume that the fixed costs are nonnegative. Using the usual terminology in location problems, a facility j is "open" when that facility is established in location j. Considering

$$y_j = \begin{cases} 1 & \text{if facility } j \text{ is open} \\ 0 & \text{otherwise} \end{cases}$$

and

$$x_{ij} = \text{fraction of the demand of client } i \text{ served by facility } j,$$

we have the so-called strong formulation of the uncapacitated facility location problem

$$\max \sum_{i \in I} \sum_{j \in J} c_{ij} x_{ij} - \sum_{j \in J} f_j y_j \qquad (UFLP)$$

$$\text{s.t.: } \sum_{j \in J} x_{ij} = 1 \quad \forall i \in I \qquad (2.1)$$

$$x_{ij} \leq y_j \quad \forall i \in I, j \in J \qquad (2.2)$$

$$y_j \in \{0,1\} \quad \forall j \in J \qquad (2.3)$$

$$x_{ij} \geq 0 \quad \forall i \in I, j \in J. \qquad (2.4)$$

Sometimes this problem appears in the literature in the minimization form

$$\min \sum_{i \in I} \sum_{j \in J} d_{ij} x_{ij} + \sum_{j \in J} f_j y_j$$

$$\text{s.t.: } (2.1), (2.2), (2.3), (2.4)$$

where d_{ij} corresponds to the transportation and service costs of satisfying the demand of client i via facility j. Observe that in the above formulation the profits associated with satisfying each client are not taken into consideration. However, these profits depend only on the client itself and by (2.1) it follows that using in the objective function the coefficients $d'_{ij} := d_{ij} - c_i$ for each $j \in J$, the value of each feasible solution is affected by $-c_i$, with c_i the profit associated with satisfying client i. Hence, adding a constant to any row of the matrix $D := [d_{ij}]$ will not change the set of optimal solutions, and as a result the two formulations are mathematically equivalent.

In this model the allocation problem can be solved trivially. Actually, in presence of a feasible set of locations, the allocation problem is solved by assigning each client to the most profitable open facility. Therefore, the decision variables correspond to the location variables y_j.

Another equivalent formulation of the uncapacitated facility location problem, the so-called weak formulation, is given by replacing constraints (2.2) by the more compact set

$$\sum_{i \in I} x_{ij} \leq m\, y_j \qquad \forall j \in J.$$

An important issue in integer programming is computational complexity. The following result classifies the uncapacitated facility location problem in the class of "difficult" problems.

Theorem 2.1.1 ([39])
The uncapacitated facility location problem is $\mathcal{NP}$-hard.

The idea behind the proof of the above theorem is to show that the $\mathcal{NP}$-hard *node packing problem* is a particular case of the uncapacitated facility location problem. The above result implies that no exact algorithm, polynomial in the number of facilities p and the number of clients m, is known. Moreover, if such an algorithm is found this would imply the existence of algorithms for all the other $\mathcal{NP}$-hard problems. Although not proved, it is assumed that no polynomial algorithm exists.

One of the basic techniques used to solve integer programming problems, in particular discrete location problems, is enumeration. The main idea behind enumeration is to divide the original problem into smaller problems, optimize them and combine these results. Carried to the extreme, enumeration will systematically list all the possible decisions. Observe that, for the uncapacitated facility location problem the total number of possible location decisions adds to 2^p. Hence, in order to prevent exhaustive enumerations, non-interesting solutions should be a priori detected and avoided during the enumeration process. A possible way to evaluate implicitly such solutions is by means of lower and upper bounds on the optimal value. Hence, it is also usual to designate an enumeration process by branch and bound, since it divides (branching) and uses bounds.

Actually, one of the first methods proposed for the uncapacitated facility location problem was the well-known heuristic due to Kuehn and Hamburger [86]. This heuristic starts by opening facilities one at a time until no additional facilities can be opened without decreasing the total profits. The criterion used to decide whether to open a new facility or not consists of determining the facility that provides the largest increase in the total profits. Having completed this first step, an improvement sequence begins. The aim is now to eliminate uneconomical facilities which were opened in earlier steps of the heuristic. This heuristic is usually known as "greedy", since at each iteration it aims only at the immediate maximum improvement. The success of the greedy heuristic is not only due to its simplicity, but also to the fact that a worst case bound for its performance can be established, see [51].

The derivation of upper bounds for integer programming problems is usually based on solving some kind of relaxation of the original problem. The simplest relaxation is probably the linear relaxation where the integrality constraints are replaced by their continuous counterpart. The quality of this relaxation depends mostly on how "close" the formulation is to the complete characterization of the convex hull of the original problem. A good example is given by the linear relaxation of the strong formulation of the uncapacitated facility location problem. In fact, the optimal solutions of this linear relaxation are often integer. However, solving this so-called strong linear relaxation by the usual linear techniques is rather inefficient due to its large dimension and intrinsic degeneracy. On the other hand, the linear relaxation of the weak formulation can easily be solved by inspection, but provides worse upper bounds. This different quality of the bounds can be explained by the fact that the feasible solution set of the strong linear relaxation is strictly contained in the feasible set of the weak linear relaxation.

Another approach to derive upper bounds is given by Lagrangian relaxation The main idea behind Lagrangian relaxation is to dualize some constraints, i.e. to weight these constraints with some multipliers and to incorporate them in the objective function. The choice of the constraints to be relaxed depends mostly on the structure of the problem to be solved, but it is usually made in such a way that the resulting relaxed problem is "easier" to solve than the original problem. For instance, Cornuéjols *et al.* [38] proposed for the uncapacitated facility location problem to relax in $(UFLP)$ the block of constraints (2.1), obtaining a Lagrangian relaxation which can be solved by inspection of the objective function. Different choices of multipliers yield different Lagrangian relaxations and thus different upper bounds. The sharpest bound is found by minimizing the associated Lagrangian relaxation over the set of multipliers, i.e. by solving the Lagrangian dual.

An alternative approach, combining basic techniques in linear programming, was proposed by Bilde and Krarup [24] and Erlenkotter [48]. This method approximates the optimal solution of the strong linear relaxation of $(UFLP)$ by considering its linear dual. Clearly, dual feasible solutions still provide upper bounds to the optimal value of $(UFLP)$. Moreover, using the complementary slackness relations it is possible to construct feasible solutions to the original problem and thus derive lower bounds. The success of the dual adjustment method lies probably on the fact that it provides at the same time good upper and lower bounds without extra effort.

Erlenkotter [48] also incorporated this idea in a branch and bound algorithm and obtained a significant improvement over the previously known results. In fact one of the first branch and bound algorithms was presented by Effroymson and Ray [46]. This branch and bound algorithm includes simplification rules to decide whether to open or close certain facilities, see Section 2.3. However, these simplification rules are not sufficiently effective to overcome the poor quality of the bounds produced by the weak linear relaxation during the enumeration process.

Other approaches to tackle this problem can be found in the extensive survey of Krarup and Pruzan [85], and in Cornuéjols *et al.* [39] included in the recent book of Mirchandani and Francis [93].

2.2 Multi-level Uncapacitated Facility Location Problems

In the uncapacitated facility location problem only one level of facilities is to be located. However, many practical situations involve more than one type of facilities and therefore multi-level models have recently received increasing attention, see [11, 17, 26, 59, 60, 81, 87, 114, 131]. For instance, Labbé and Wendell [87] discuss the case of a company specialized in industrial garbage treatment which needs to find the optimal location of truck depots and plants for disposal of garbage.

In this section we will concentrate on models where more than one type of facility is to be simultaneously located. However, we will not consider models which take into account extra constraints relating the different types of facilities, like the ones presented in [81, 114]. Kaufman *et al.* [81] propose a model in which facilities and depots have to be located simultaneously assuming that a facility can only be opened if its associated depot is open. Ro and Tcha [114] extended the above model by considering two types of facilities, some of which can be opened independently of the depots while others can only be opened if all their associated depots are open.

2.2.1 The Multi-level Uncapacitated Facility Location Problem

Tcha and Lee [131] consider the *multi-level uncapacitated facility location problem* where more than one type of facilities has to be simultaneously located. This model is particularly suited for situations where products have to be shipped from origin points or supply levels, to the demand points or clients via intermediate level facilities. Thus, the objective is to choose where to simultaneously locate facilities in each level in order to maximize the profit associated with satisfying the demand of a given set of clients minus the fixed costs of opening facilities in each level. Again, no limitations on the capacity of either of the facility types are considered.

Let $I := \{1, \ldots, m\}$ denote the set of clients, l the number of facility levels and $J_r := \{1, \ldots, p_r\}$ the set of sites where facilities can be located in level r. Consider P the set of all possible paths from facilities in the first level to the last level l, i.e. $P := \{(j_1 \ldots j_l) : j_r \in J_r, r = 1, \ldots, l\}$. Denote by $P(j_r) \subseteq P$ the set of all possible paths that include facility j_r, and by f_{j_r} the fixed costs associated with facility j_r.

Let c_{iw} be the profit associated with satisfying the demand of client i via the path $w \in P$. Considering, for each level r,

$$y_{j_r} = \begin{cases} 1 & \text{if facility } j \text{ in level } r \text{ is open} \\ 0 & \text{otherwise} \end{cases}$$

and x_{iw} = fraction of the demand of client i served by the path $w = (j_1 \ldots, j_l)$, Tcha and Lee [131] propose the following formulation for the multi-level uncapacitated facility location problem

$$\max \sum_{i \in I} \sum_{w \in P} c_{iw} x_{iw} - \sum_{r=1}^{l} \sum_{j_r \in J_r} f_{j_r} y_{j_r} \qquad (MUFLP)$$

$$\text{s.t.:} \sum_{w \in P} x_{iw} = 1 \quad \forall i \in I \qquad (2.5)$$

$$x_{iw} \leq y_{j_r} \quad \forall i \in I, w \in P(j_r), j_r \in J_r, r = 1, \ldots, l \qquad (2.6)$$

$$y_{j_r} \in \{0, 1\} \quad \forall j_r \in J_r, r = 1, \ldots, l \qquad (2.7)$$

$$x_{iw} \geq 0 \quad \forall i \in I, w \in P, r = 1, \ldots, l. \qquad (2.8)$$

If there are no fixed costs of opening facilities in each level, this model reduces to the uncapacitated facility location problem. This clearly implies that the $(MUFLP)$ is $\mathcal{NP}$-hard, since it is a generalization of an $\mathcal{NP}$-hard problem, see [61].

Observe that formulation $(MUFLP)$ corresponds to the extension to the multi-level case of the strong formulation of the uncapacitated facility location problem. Tcha and Lee [131] also mention that constraints (2.6) can be replaced by the following smaller group of constraints

$$x_{iw} \leq \frac{\sum_{j_r \in w} y_{j_r}}{l} \qquad \forall w \in P, i \in I.$$

However, according to these authors the bounds provided by the linear relaxation of this more compact formulation are not as good as the ones generated by the linear relaxation of $(MUFLP)$. In order to approximate the optimal value of the linear relaxation of $(MUFLP)$, Tcha and Lee [131] present an extension of the dual adjustment method of Erlenkotter [48]. A primal ascent method which tries to improve the primal solution given by the dual descent method is also given. These two bounding methods are combined in a binary branch and bound algorithm on the variables $y_{j j_r}$. However, this branch and bound algorithm also includes an incorrect node simplification method, see Section 2.3, and therefore the computational results presented in [131] may not be correct.

Lately, the interest on multi-level uncapacitated facility location problems has shifted towards the case where only two levels of facilities exist, i.e. the 2-level uncapacitated facility location problem, see [1, 17]. In particular, the general model for the 2-level case proposed by Barros and Labbé [17] and discussed in detail in Section 2.4 includes the 2-level uncapacitated facility location problem as a special case. As shown in this section, the linear relaxation bounds produced by $(MUFLP)$ for $p = 2$ can be dominated by another "stronger" formulation of the problem. The computational results presented in Section 2.4.7 show that in practice the duality gaps for this problem are considerably large and very dependent on the type of formulation used. Recently, Aardal *et al.* [1] have been investigating valid inequalities to improve this "stronger" formulation.

2.2.2 The 2-echelon Uncapacitated Facility Location Problem

The 2-*level uncapacitated facility location problem*, where only two levels of facilities are considered, can be related to the 2-*echelon uncapacitated facility location problem*, proposed by Gao and Robinson [59]. In this model the products also have to be shipped from one level, echelon-1, to the clients via an intermediate second level of facilities, echelon-2, before reaching the client. The fixed costs induced by facilities in echelon-2 depend not only on the location but also on the facilities in echelon-1 that supply the products. Thus, instead of considering a fixed cost associated to each open facility in echelon-2, Gao and Robinson [59] assume that a fixed cost is associated to each pair of echelon-1 and echelon-2 facilities that serves at least one client, i.e. operates together. Again, no limitations on the capacity of either of the facility types are imposed. Applications of this model can be found, for instance, in distributed computer networks and postal collection, see [59].

Gao and Robinson [59] described this problem in the following way. The echelon-1 type facilities correspond to distribution centers that supply the echelon-2 type facilities, i.e. the depots. Let $I := \{1, \ldots, m\}$ denote the set of clients, $J := \{1, \ldots, p\}$ the set of sites where distribution centers can be located and $K := \{1, \ldots, q\}$ the set of sites where depots can be located. The fixed cost of opening the distribution center j is denoted by f_j and the fixed cost associated with depot k and distribution center j is given by F_{jk}. Finally, the profits associated with satisfying client i via the distribution center j and depot k are represented by c_{ijk}. Considering,

$$y_j = \begin{cases} 1 & \text{if distribution center } j \text{ is open} \\ 0 & \text{otherwise} \end{cases},$$

$$t_{jk} = \begin{cases} 1 & \text{if depot } k \text{ is open and supplied by distribution center } j \\ 0 & \text{otherwise} \end{cases}$$

and x_{ijk} = fraction of the demand of client i served through distribution center j and depot k, the 2-echelon uncapacitated facility location problem can be formulated as

$$\max \sum_{i \in I} \sum_{j \in J} \sum_{k \in K} c_{ijk} x_{ijk} - \sum_{j \in J} f_j y_j - \sum_{j \in J} \sum_{k \in K} F_{jk} t_{jk} \qquad (2ELP)$$

$$\text{s.t.:} \quad \sum_{j \in J} \sum_{k \in K} x_{ijk} = 1 \qquad \forall i \in I \qquad (2.9)$$

$$x_{ijk} \leq t_{jk} \qquad \forall i \in I, j \in J, k \in K \qquad (2.10)$$

$$t_{jk} \leq y_j \qquad \forall j \in J, k \in K \qquad (2.11)$$

$$y_j, t_{jk} \in \{0,1\} \qquad \forall j \in J, k \in K \qquad (2.12)$$

$$x_{ijk} \geq 0 \qquad \forall i \in I, j \in J, k \in K. \qquad (2.13)$$

Clearly, the uncapacitated facility location problem is a particular case of this problem and therefore $(2ELP)$ is also $\mathcal{NP}$-hard, see [61].

Observe that the above formulation corresponds to the extension to the 2-level case of the strong formulation of the uncapacitated facility location problem. In a similar way, Gao and Robinson [59] mention that constraints (2.10) and (2.11) can be replaced by the following smaller group of constraints

$$\sum_{i \in I} x_{ijk} \leq m\, t_{jk} \qquad \forall j \in J, k \in K$$

$$\sum_{k \in K} t_{jk} \leq q\, y_j \qquad \forall j \in J$$

which yields an equivalent formulation. However, according to [59], the bounds provided by the linear relaxation of this condensed formulation are weaker than the ones obtained by using the linear relaxation of $(2ELP)$.

Gao and Robinson [59] extend the dual adjustment method of Erlenkotter to the $(2ELP)$ formulation. They also provide a primal ascent method which tries to improve the primal solution given by the dual descent method. In [59] computational results on a binary branch and bound algorithm on the variables y_j and t_{jk} and using these two bounding methods, are reported. From these results it appears that the linear relaxation of $(2ELP)$ is quite strong.

The 2-echelon uncapacitated facility location problem is a special case of the general model proposed by Barros and Labbé [17] and discussed in detail in Section 2.4. As we shall see in Section 2.4, the linear relaxation bounds produced by the basic formulation $(2ELP)$ can be improved by considering a "stronger" formulation. This is confirmed by the computational results reported in Section 2.4.7.

As mentioned before, the 2-echelon uncapacitated facility location problem generalizes the classical uncapacitated facility location problem. However, as shown by Gao and Robinson [59, 60], the 2-echelon uncapacitated facility location problem also generalizes the *multi-activity uncapacitated facility location problem* introduced by Klincewicz *et al.* [84]. This problem corresponds to a 1-level location problem where the clients require different products (activities). Therefore, the objective is to find how many and which facilities should be opened and also which products will be handled by the open facilities. Clearly, the fixed costs in this model correspond to the usual fixed costs of opening facilities and the fixed costs of equipping facilities to handle certain products. Finally, the profits reflect the gains obtained by supplying each client with a different product via a given facility. Although this problem is in essence a 1-level location problem, it can be translated into a 2-echelon uncapacitated facility location problem. This is achieved by creating as many replicas of each client as the available products and by considering the different products as 2-echelon facilities. This "other" interpretation of the multi-activity uncapacitated facility location problem provides an alternative way to solve this problem using the general solution techniques for the 2-echelon uncapacitated facility location problem. However, a multi-activity uncapacitated facility location problem yields a larger 2-echelon uncapacitated facility location problem and this increase in dimension is reflected in the computational time required for solving these transformed problems, see [60].

2.3 Submodularity

An important issue for discrete location problems is to know if some modularity property is satisfied. Thus, we will start by recalling two equivalent definitions of submodularity.

Definition 2.3.1 ([97, 96])
Let N be a finite set and let Z a real-valued function defined on the subsets of N. The function Z is submodular if

$$Z(R \cup \{t\}) - Z(R) \geq Z(S \cup \{t\}) - Z(S)$$

for all $R \subseteq S \subseteq N$ and $t \notin S$.

On the other hand, a function Z is supermodular if $-Z$ is submodular. Moreover, if a function Z is simultaneously submodular and supermodular then it is modular.

The submodularity concept can also be expressed as follows.

Proposition 2.3.1 ([97])
Let N be a finite set and let Z a real-valued function defined on the subsets of N. Defining $\rho_S(r) := Z(S \setminus \{r\}) - Z(S)$, Z is submodular if and only if

$$\rho_{S'}(r) \leq \rho_S(r)$$

for all $S' \subseteq S \subseteq N$ and $r \in S'$.

Even though there exists a polynomial algorithm for the minimization of a submodular function, the problem of maximizing a submodular function is $\mathcal{NP}$-hard, see [58, 96]. However, some important properties can be derived for the maximization of a submodular function. Namely, it is well-known that the greedy heuristic has a worst case guaranteed performance for the maximization of submodular functions, see [97]. Furthermore, this property allows the design of some useful simplification rules that will reduce the size of branch and bound trees, when solving the problem to optimality.

Since the location models discussed are considered in the maximization form, we will concentrate on the submodularity property in this section. Clearly, if the models are proposed in the minimization form the relevant property is supermodularity, see [7, 57]. A well-known example of a submodular problem is given by the uncapacitated facility location problem, which can be formulated as follows

$$\max_{S \subseteq J} Z(S)$$

with $Z(S) := \sum_{i \in I} \max_{j \in S} c_{ij} - \sum_{j \in S} f_j$ and J representing the set of possible locations and I the set of clients. The greedy heuristic presented by Kuehn and Hamburger [86] is not only "easy" to use in practice but it also has an important theoretical property. In fact, submodularity permits to derive a worst case bound for the performance of the greedy heuristic, see [51]. Moreover, also node simplifications can easily be derived for this problem, see [57]. An example of a simplification rule is the following rule to fix facilities open in a branch and a bound method. Let S denote the set of non closed facilities in some node of the branch and bound tree. Due to Proposition 2.3.1, a free facility r can be fixed open in all descendant nodes of S as soon as $\rho_S(r) < 0$. Similarly simplification rules to close facilities can also be derived. Observe that the simplification rules proposed by Efroymson and Ray [46] for the uncapacitated facility location problem in the minimization form can be justified using results on supermodular functions.

The above remarks suggest to investigate if the general multi-level uncapacitated facility location problem also satisfies the submodularity property. Let J_r for each $r = 1, \ldots, l$ denote the set of sites where facilities of level r can be selected. To each facility j_r, a fixed cost f_{j_r} of opening this facility is associated. Let $I := \{1, \ldots, m\}$ denote the set of clients and the profit associated with serving client i from the l facilities $j_1, \ldots, j_l$ by $c_{ij_1\ldots j_l}$. For each given subset $S_r \subseteq J_r$, $r = 1, \ldots, l$, of open facilities, the corresponding solution is given by assigning each client to the most profitable combination of l open facilities. Hence, the multi-level uncapacitated facility location problem can be formulated as follows

$$\max_{S \subseteq J} Z(S)$$

with $Z(S) := \sum_{i \in I} \max_{j_1 \in S_1, \ldots, j_l \in S_l} c_{ij_1 \ldots j_l} - \sum_{r=1}^{l} \sum_{j_r \in S_r} f_{j_r}$ and $J := \bigcup_{r=1}^{l} J_r$ and $S := \bigcup_{r=1}^{l} S_r$. This means that the multi-level uncapacitated facility location problem reduces to the maximization of a real-valued function defined on the subsets of J.

Consider now the following example of a 2-level uncapacitated facility location problem.

Example 2.3.1

Let the set of the facilities in the first level be given by $J_1 := \{1_1, 2_1\}$ and the set of facilities in the second level be given by $J_2 := \{1_2, 2_2\}$. Assume that all the fixed costs are zero and let the profits associated with the only client be given as follows

$c_{1..}$	1_2	2_2
1_1	1	1
2_1	100	1

Take $R \subset S$ defined as $R := R_1 \cup R_2 := \{1_1\} \cup \{2_2\}$ where $R_1 \subseteq J_1$ and $R_2 \subseteq J_2$ and $S := S_1 \cup S_2 := \{1_1\} \cup \{1_2, 2_2\}$ with $S_1 \subseteq J_1$ and $S_2 \subseteq J_2$. Hence, $Z(R) = Z(S) = 1$. Assume now that facility 2_1 is also open. We get $Z(R \cup \{2_1\}) = 1$ and $Z(S \cup \{2_1\}) = 100$ which contradicts Definition 2.3.1. Hence, the multi-level uncapacitated facility location problem, for $l > 1$ is not submodular, although its special case for $l = 1$, the uncapacitated facility location problem, is submodular.

It is also interesting to check whether the existence of side constraints in the model will have any effect in the presence or not of the submodularity property. Therefore, we will consider the model proposed by Ro and Tcha [114], i.e. a 2-level uncapacitated facility location problem with special relations between facilities on different levels. These relations correspond to only opening some facilities in the first level if all the associated facilities in the second level are open. In order to create an equivalent instance of this problem in the maximization form, we will consider Example 2.3.1 with the additional constraint that facility 1_1 in the first level can only be opened if facility 2_2 in the second level is open. It is easy to verify that neither the submodular nor the supermodularity property hold. In spite of this, the simplification rules presented by Ro and Tcha [114] are still valid, since they are based on lower bounds on the reduced costs $\rho_S(r)$ and thus are not direct consequences of supermodularity, see [16].

On the other hand, Tcha and Lee [131] assume that the multi-level uncapacitated facility location problem is submodular (supermodular in the minimization case). However, since this assumption does not hold, their node simplification rule to fix facilities open in the branch and bound algorithm is not valid. We will consider the maximization form of the problem and show that, unfortunately, the equivalent rule to fix facilities open is wrong. Let S denote the set of non closed facilities in some node of the branch and bound tree and let S' be the set of non closed facilities corresponding to a descendant node. Clearly, $S' \subseteq S$. Now, by Proposition 2.3.1, $\rho_{S'}(r) \leq \rho_S(r)$ for r in the set of the free facilities (for all S', S, r with $S' \subseteq S$ and $r \in S'$) if and only if f is submodular. The simplification rule for the maximization form consists in opening a facility as soon as $\rho_S(r) \leq 0$. This rule is not valid since the monotonicity of $\rho_S(r)$ cannot be guaranteed (the same holds for the original rule in the minimization case of Tcha and Lee [131]). The following example shows clearly why the rule is not valid.

Example 2.3.2

Let the set of the facilities in the first level be given by $J_1 := \{1_1, 2_1\}$ and the set of facilities in the second level be $J_2 := \{1_2, 2_2\}$. Consider also the fixed costs $f_{1_1} = f_{1_2} = 0$ and $f_{2_1} = f_{2_2} = 9$. The profits associated with the only client are given by

$c_{1..}$	1_2	2_2
1_1	20	10
2_1	10	30

Assume that we are at a node ℓ of the branch and bound tree for which facilities 1_1 and 1_2 are open and all the others are free. According to this rule, we would immediately open both facilities 2_1 and 2_2 since $\rho(2_1) = Z(\{1_1, 1_2, 2_2\}) - Z(\{1_1, 1_2, 2_1, 2_2\}) = 20 - (30 - f_{2_1}) = -1 = \rho(2_2)$ and stop the enumeration from this node with a solution of value 12. However, a better solution can be found in descendant nodes of ℓ. In fact, we simply need to consider the descendant node where both 2_1 and 2_2 are closed (fixed to zero) to obtain a solution of value 20.

Similarly, Example 2.3.2 with $f_{2_1} = f_{2_2} = 1$ can be used to show that the corresponding rule to fix facilities closed is also not valid. In this case the rule consists in fixing a free variable r to be closed if $\rho_{S \cup \{r\}}(r) \geq 0$, where S is now the set of fixed open facilities. A valid branch and bound algorithm will thus involve much larger branch and bound trees and take much more time to reach optimality, see [17].

Hence, submodularity does not hold for this class of location problems where more than one type of facilities has to be simultaneously located, independently of how the associated fixed costs are generated.

2.4　A General Uncapacitated Facility Depot Location Model

In this section we will discuss in detail the general model proposed by Barros and Labbé [17]. This model further generalizes the 2-level and 2-echelon uncapacitated facility location problems in terms of the fixed costs that are allowed. Three different formulations are proposed for this general model in Section 2.4.2 and their linear relaxations are discussed in Section 2.4.3. Lagrangian relaxation techniques are briefly discussed in Section 2.4.4 and a modified subgradient is derived to solve special Lagrangian duals. Finally, these techniques are applied to the formulations presented in Section 2.4.2 in order to generate upper bounds on the optimal value of the problem. Heuristics to obtain lower bounds for this problem are described in Section 2.4.5 while a branch and bound algorithm is presented in Section 2.4.6. Computational results using the approaches discussed in the previous sections are presented and discussed in Section 2.4.7. In conclusion, some final remarks are given in Section 2.5.

2.4.1　Introduction

We will consider a general model for the uncapacitated facility and depot location problem of locating facilities (first level) and depots (second level) in order to maximize the sum of the profits associated with serving each client by exactly one facility and one depot, taking into account the fixed costs of opening those facilities and depots and also the fixed costs of having facilities and depots operating together. Thus, the costs associated with having certain entities, for instance facilities and depots, open and serving a client are given not only by the fixed costs of opening these entities but also by the infrastructure (operating) costs of having certain pairs of entities (facility, depot) operating together. Observe that all these fixed costs are present in either the 2-level or in the 2-echelon uncapacitated facility location problems but they have never been considered simultaneously.

A possible application of this general model, among others, is the location of truck depots and plants for disposal of industrial garbage (facilities). Labbé and Wendell [87] discuss the case of a company specialized in industrial garbage treatment. The company must decide the number and location of both the truck depots and the plants for disposal of garbage in order to maximize its profit. In order to satisfy the demand of each client, a truck leaves a depot, travels to the client, collects the garbage, brings it to the disposal plant and returns to the depot. In determining the profit of such a service, the travel cost of the triangular tour made to satisfy the demand of the client is also taken into account, see Figure 2.2. An example of such a profit function is given by $c_{ijk} := d_i \left(p_i - q_j - r_k - t\, D_{ijk} \right)$, where d_i is the demand of client i, p_i the profit per unit served to client i, q_j the production cost in facility j per unit, r_k the transportation cost per unit by a truck stationed in depot k, t the travel cost per unit and kilometer and D_{ijk} the length of the triangular tour $\text{depot}_k \rightarrow \text{client}_i \rightarrow \text{facility}_j \rightarrow \text{depot}_k$.

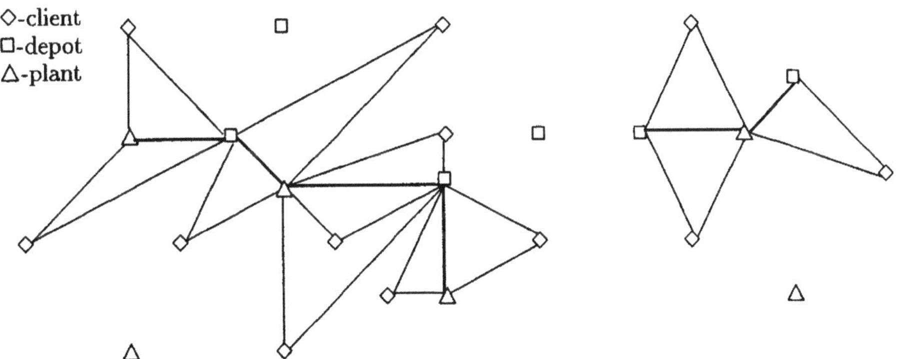

Figure 2.2: An example of a solution to the general uncapacitated facility and depot location problem.

In the situation described above it may not be possible to have each combination of depot and disposal plant operating together, since disposal plants need to be adequately equipped to handle trucks from particular depots. Hence, it is assumed that an investment can be made to make disposal plants suitable for receiving trucks from particular depots. Obviously, investments designed to improve the infrastructures are fixed costs in our model, and hence the term "operating (infrastructure) costs" was adopted. Such costs are associated with the operating pairs, that are illustrated in Figure 2.2 with darker arcs. Examples of other investments that can be considered in this sense are costs to improve or even construct tunnels, roads and bridges, linking the two different types of facilities.

Clearly, if there are no fixed costs associated with having facilities and depots operating together and there are also no fixed costs of opening depots, then this problem reduces to the well-known uncapacitated facility location problem. This immediately implies that this problem is $\mathcal{NP}$-hard, since it is a generalization of an $\mathcal{NP}$-hard problem, see [61]. On the other hand, when there are no fixed costs associated with having facilities and depots operating together, the general uncapacitated facility and depot location problem reduces to the multi-level uncapacitated facility location problem as defined by Tcha and Lee [131] with two levels.

Finally, if there are no fixed costs associated with opening depots, then the model reduces to that proposed by Gao and Robinson [59]. However, our model is not suitable for the situations where special relations between the depots and facilities exist, such as when facilities can only be opened if a prescribed depot is open, see Kaufman *et al.* [81] and Ro and Tcha [114].

2.4.2 Formulations

We consider a set of clients $I := \{1, \ldots, m\}$, a set of sites $J := \{1, \ldots, p\}$ where facilities can be located, and a set of sites $K := \{1, \ldots, q\}$ where depots can be located. The fixed cost of opening facility j is f_j and the fixed cost of opening depot k is g_k. The fixed cost of having facility j and depot k operating together is F_{jk}. We will assume that all the fixed costs are nonnegative. Let c_{ijk} denote the profit associated with satisfying the demand of client i from facility j and depot k. We want to determine which facilities and depots should be opened, and which should operate together in order to maximize the total profit, given that the demand of all clients is satisfied.

Consider the profits c_{ijk} and the costs f_j, g_k and F_{jk} as defined above and let

$$y_j = \begin{cases} 1 & \text{if facility } j \text{ is open} \\ 0 & \text{otherwise} \end{cases},$$

$$z_k = \begin{cases} 1 & \text{if depot } k \text{ is open} \\ 0 & \text{otherwise} \end{cases},$$

$$t_{jk} = \begin{cases} 1 & \text{if } j \text{ and } k \text{ are operating together} \\ 0 & \text{otherwise} \end{cases}$$

and $x_{ijk} =$ fraction of the demand of client i served by the operating pair (j, k).

We first consider the following (Weak) Formulation for the problem

$$\max \sum_{i \in I} \sum_{j \in J} \sum_{k \in K} c_{ijk} x_{ijk} - \sum_{j \in J} f_j y_j - \sum_{k \in K} g_k z_k - \sum_{j \in J} \sum_{k \in K} F_{jk} t_{jk} \qquad (WF)$$

$$\text{s.t.:} \ \sum_{j \in J} \sum_{k \in K} x_{ijk} = 1 \quad \forall i \in I \qquad (2.14)$$

$$x_{ijk} \leq t_{jk} \qquad \forall i \in I, j \in J, k \in K \qquad (2.15)$$

$$t_{jk} \leq y_j \qquad \forall j \in J, k \in K \qquad (2.16)$$

$$t_{jk} \leq z_k \qquad \forall j \in J, k \in K \qquad (2.17)$$

$$y_j, z_k \in \{0, 1\} \qquad \forall j \in J, k \in K \qquad (2.18)$$

$$t_{jk} \in \{0, 1\} \qquad \forall j \in J, k \in K \qquad (2.19)$$

$$x_{ijk} \geq 0 \qquad \forall i \in I, j \in J, k \in K. \qquad (2.20)$$

Constraints (2.14) guarantee that the demand of every client is satisfied, constraints (2.15) assure that clients are only supplied from (facility,depot) pairs, which are operating together and constraints (2.16) and (2.17) guarantee that an operating pair has both of its components open.

If the following $m(p + q)$ constraints are added to formulation (WF)

$$\sum_{k \in K} x_{ijk} \leq y_j \quad \forall i \in I, j \in J, \qquad (2.21)$$

$$\sum_{j \in J} x_{ijk} \leq z_k \quad \forall i \in I, k \in K, \qquad (2.22)$$

we will obtain the equivalent (Big Strong) Formulation (BSF) for our problem.

Moreover, due to the integrality of the variables y_j, z_k and t_{jk} and the observation that the objective function has nonnegative coefficients associated with the variables

t_{jk}, it follows that constraints (2.16) and (2.17) can be removed from (BSF), yielding a smaller equivalent (Strong) Formulation (SF).

It is also interesting to point out the modifications in the above formulations for the special cases: uncapacitated facility location problem, 2-level uncapacitated facility location problem, and 2-echelon uncapacitated facility location problem.

Clearly, when $F_{jk} = 0$ for all $j \in J, k \in K$ and $g_k = 0$ for all $k \in K$ the model reduces to the well-known uncapacitated facility location problem. In this case, after simplifying formulation (WF) we obtain the well-known strong formulation $(UFLP)$.

The 2-level uncapacitated facility location problem corresponds to the case where $F_{jk} = 0$ for all $j \in J$ and $k \in K$. Clearly, all the three formulations presented can be simplified by removing variables t_{jk}. In particular, the formulation proposed by Tcha and Lee [131] corresponds to the simplified (WF).

Finally, the 2-echelon uncapacitated facility location problem can be derived by taking $g_k = 0$ for all $k \in K$. Again, the presented formulations can be simplified by removing variables z_k. Moreover, the formulation proposed by Gao and Robinson [59, 60] corresponds to the simplified (WF).

2.4.3   Linear Relaxation

Linear relaxation is probably one of the most popular approaches to derive upper bounds on the optimal value of a (mixed) integer problem. As exemplified for the uncapacitated facility location problem, equivalent formulations may yield distinct linear relaxations, and thus qualitatively different upper bounds. Moreover, it appears that the trade off between the computational effort spent in computing these bounds and their quality is a significant aspect to take into consideration. Therefore, it is important to relate the linear relaxations of the three formulations presented in the previous section.

For simplicity, we denote by $\vartheta(P)$ the optimal value of a problem (P) and by $(\overline{P})$ its linear relaxation.

Proposition 2.4.1

$$\vartheta(WF) = \vartheta(BSF) = \vartheta(SF) \leq \vartheta(\overline{SF}) = \vartheta(\overline{BSF}) \leq \vartheta(\overline{WF}).$$

Proof: The first three equalities follow immediately from the formulations (WF), (BSF) and (SF) and the first inequality from the definition of linear relaxation.

It is easy to see that any feasible solution to $(\overline{BSF})$ is also feasible to $(\overline{WF})$ and therefore $\vartheta(\overline{BSF}) \leq \vartheta(\overline{WF})$. Furthermore, any feasible solution to $(\overline{BSF})$ is also feasible to $(\overline{SF})$, and thus $\vartheta(\overline{BSF}) \leq \vartheta(\overline{SF})$. Now let $(\bar{t}, \bar{y}, \bar{z}, \bar{x})$ be an optimal solution to $(\overline{SF})$. Since all coefficients F_{jk} are nonnegative, we can assume due to constraints (2.15) that

$$\bar{t}_{jk} = \bar{x}_{i^*jk} := \max_{i \in I} \bar{x}_{ijk} \quad \forall j \in J, k \in K. \tag{2.23}$$

Furthermore, since $\bar{x}_{i^*jk} \leq \sum_{k \in K} \bar{x}_{i^*jk}$ (respectively $\bar{x}_{i^*jk} \leq \sum_{j \in J} \bar{x}_{i^*jk}$) it follows that (2.23) and (2.21) (respectively (2.22)) imply (2.16) (respectively (2.17)). As a consequence, $(\bar{t}, \bar{y}, \bar{z}, \bar{x})$ is also feasible to $(\overline{BSF})$ and so $\vartheta(\overline{SF}) \leq \vartheta(\overline{BSF})$. $\qquad\square$

As will be confirmed by our computational results, $(\overline{WF})$ provides a significantly worse upper bound on the optimal value of the problem than $(\overline{SF})$ or $(\overline{BSF})$. This phenomenon may be explained by the fact that $(\overline{WF})$ allows some fractional feasible solutions that are forbidden in $(\overline{SF})$ and $(\overline{BSF})$. For $p > 1$ or $q > 1$, the solution $x_{ijk} = t_{jk} = y_j = z_k = \frac{1}{pq}$ is a typical example. Moreover, it can easily be shown that (2.21) and (2.22) define facets of the convex hull of the feasible solutions of this problem, see [1].

Although $(\overline{SF})$ and $(\overline{BSF})$ provide the same bound, their feasible sets do not coincide. In fact, if $p > 1$ or $q > 1$, the solution $x_{ijk} = y_j = z_k = \frac{1}{pq}$ and $t_{jk} = 1$ is feasible to $(\overline{SF})$ but not to $(\overline{BSF})$. However, this type of solution is not interesting due to (2.23). Hence, in spite of being more restrictive than $(\overline{SF})$, $(\overline{BSF})$ does not improve the linear bound as shown by Proposition 2.4.1.

Due to the large number of constraints and variables, solving $(\overline{WF})$ or $(\overline{SF})$ requires a powerful linear solver. Moreover, these linear programming problems are very often highly degenerate and therefore it does not seem advisable to solve them directly. Even though the approach followed here does not consider solving directly these linear relaxations or their duals, it is nevertheless interesting to state these linear duals. Furthermore, some of the results reported in the next section will use the knowledge of the dual feasible set.

Observe first that the upper bounding constraints on variables y_j, z_k and t_{jk} can be dropped in the three linear relaxations. We will begin by considering the linear dual of $(\overline{BSF})$

$$\min \sum_{i \in I} \lambda_i \qquad\qquad (LDBSF)$$

$$\text{s.t.:} \quad \lambda_i + u_{ijk} + \alpha_{ij} + \beta_{ik} \geq c_{ijk} \quad \forall i \in I, j \in J, k \in K \qquad (2.24)$$

$$-\sum_{i \in I} u_{ijk} + v_{jk} + w_{jk} \geq -F_{jk} \quad \forall j \in J, k \in K \qquad (2.25)$$

$$-\sum_{k \in K} v_{jk} - \sum_{i \in I} \alpha_{ij} \geq -f_j \quad \forall j \in J \qquad (2.26)$$

$$-\sum_{j \in J} w_{jk} - \sum_{i \in I} \beta_{ik} \geq -g_k \quad \forall k \in K \qquad (2.27)$$

$$u_{ijk}, v_{jk}, w_{jk}, \alpha_{ij}, \beta_{ik} \geq 0 \quad \forall i \in I, j \in J, k \in K. \qquad (2.28)$$

From the above constraints it is possible to establish limits on the variation of the dual variables

$$\max_{j \in J, k \in K} (c_{ijk} - F_{jk} - g_k - f_j) \leq \lambda_i \leq \max_{j \in J, k \in K} c_{ijk} \qquad (2.29)$$

$$0 \leq u_{ijk} \leq F_{jk} + g_k + f_j \qquad (2.30)$$

$$0 \leq v_{jk} \leq f_j \qquad (2.31)$$

$$0 \leq w_{jk} \leq g_k \qquad (2.32)$$

$$0 \leq \alpha_{ij} \leq f_j \qquad (2.33)$$

$$0 \leq \beta_{ik} \leq g_k. \qquad (2.34)$$

The utility of these bounds will become clear in the next section. However, they can be used to establish a trivial upper bound on the optimal value of $(\overline{BSF})$

$$\sum_{i \in I} \max_{j \in J, k \in K} (c_{ijk} - F_{jk} - g_k - f_j).$$

Since $(\overline{SF})$ differs from $(\overline{BSF})$ only on the sets of constraints (2.16) and (2.17), it is enough to remove from $(LDBSF)$ the dual variables v_{jk} and w_{jk} to obtain the dual of $(\overline{SF})$. Clearly, (2.29), (2.30), (2.33) and (2.34) still establish valid limits on the variation of the dual variables.

In a similar way, the linear dual of $(\overline{WF})$ can be obtained by eliminating from $(LDBSF)$ the variables α_{ij} and β_{ik}. For this dual (2.29), (2.30), (2.31) and (2.32) also establish valid limits on the variation of the dual variables.

2.4.4   Lagrangian Relaxation

Many "difficult" (mixed) integer programming problems can be formulated as a relatively easy problem complicated by a set of hard constraints. For these cases, Lagrangian relaxation is a particularly adequate tool to derive bounds for the original problem. One of the most popular examples of how Lagrangian relaxation can be applied to integer programming is given by the pioneering work of Held and Karp [71]. Although this technique is a special case of the more general framework of Lagrangian duality in nonlinear optimization, we will only review it in the operations research perspective. Briefly, a Lagrangian relaxation corresponds to a problem in which the complicating constraints are replaced by a penalty term in the objective function. This penalty term is a function of the amount of violation of these constraints weighted by multipliers. In order to exemplify this technique consider the following (mixed) integer programming problem

$$\max\left\{ c^{\mathsf{T}} x : Ax \leq b, Bx \leq d, x \geq 0, x_j \text{ integer for } j = 1, \ldots, n \right\} \qquad (IP)$$

where $c \in \mathbb{R}^n$, $b \in \mathbb{R}^m$, $d \in \mathbb{R}^p$ and $A \in \mathbb{R}^{m \times n}$, $B \in \mathbb{R}^{p \times n}$. Assume that this problem could easily be solved if the set of complicating constraints $Ax \leq b$ would be removed. Hence, the natural Lagrangian relaxation of (IP) is to associate the set of multipliers $\lambda \geq 0$ to these constraints, yielding the problem

$$\mathcal{L}^{IP}(\lambda) := \max\left\{ c^{\mathsf{T}} x + \lambda^{\mathsf{T}}(b - Ax) : Bx \leq d, x \geq 0, x_j \text{ integer for } j \in J \right\}. \quad (\mathcal{L}^{IP}_\lambda)$$

Clearly, any feasible solution to (IP) is feasible to $(\mathcal{L}^{IP}_\lambda)$. Moreover, for any $\lambda \geq 0$, the associated Lagrangian relaxation yields a valid upper bound on the optimal value of (IP). In order to find the *sharpest* bound, the following convex programming problem has to be solved

$$\min\left\{ \mathcal{L}^{IP}(\lambda) : \lambda \geq 0 \right\}. \qquad (\mathcal{D}\mathcal{L}^{IP})$$

This problem is the Lagrangian dual of (IP) with respect to the constraint set $Ax \leq b$. Ideally, the optimal value of the above problem should coincide with the optimal value of (IP). However, in most integer programming problems this is not the case, and therefore a duality gap exists. This duality gap is usually measured by taking the relative difference of the optimal values. An important issue when considering upper bounding methods is the size of the gap that can be expected. In particular, what is the relation between the bounds derived using a Lagrangian dual and the linear relaxation? In order to analyze this relation, we will start by considering the following relaxation of (IP)

$$\max c^{\mathsf{T}} x \qquad\qquad (IP^\star)$$

$$\text{s.t.:}\ Ax \leq b$$

$$x \in \mathrm{co}\{x \geq 0 : Bx \leq d, x_j \text{ integer for } j \in J\}$$

where co represents the convex hull of a set. Although difficult to describe, the convex hull of a set can always be defined using linear constraints and therefore $(IP^\star)$ is a linear programming problem. Also an important concept in Lagrangian duality is the *integrality property*.

Definition 2.4.1

A Lagrangian relaxation satisfies the integrality property whenever

$$\vartheta(\mathcal{L}_\lambda^{IP}) = \vartheta(\overline{\mathcal{L}_\lambda^{IP}}) \text{ for all } \lambda \geq 0.$$

The following theorem condenses the relations between these different relaxations.

Theorem 2.4.1 ([62])

1. $\vartheta(IP) \leq \vartheta(IP^\star) \leq \vartheta(\overline{IP})$ and $\vartheta(IP) \leq \vartheta(\mathcal{L}_\lambda^{IP})$ for any $\lambda \geq 0$.

2. If for a given λ a vector x satisfies the three following conditions

    (i) x is optimal to $(\mathcal{L}_\lambda^{IP})$,

    (ii) $Ax \leq b$,

    (iii) $\lambda^{\mathsf{T}}(b - Ax) = 0$,

then $\boldsymbol{x}$ is an optimal solution to (IP).

If $\boldsymbol{x}$ satisfies (i) and (ii) but not (iii) then $\boldsymbol{x}$ is an ε-optimal solution to (IP) with $\varepsilon := \boldsymbol{\lambda}^{\mathsf{T}}(\boldsymbol{b} - \boldsymbol{Ax})$.

3. *If $(\overline{IP})$ is feasible then $\vartheta(\mathcal{L}_{\overline{\boldsymbol{\lambda}}}^{IP}) \leq \vartheta(\overline{IP})$ for $\overline{\boldsymbol{\lambda}}$ an optimal dual vector associated with constraints $\boldsymbol{Ax} \leq \boldsymbol{b}$.*

4. *If $(IP^{\star})$ is feasible then $\vartheta(\mathcal{DL}^{IP}) = \vartheta(\mathcal{L}_{\boldsymbol{\lambda}^{\star}}^{IP}) = \mathcal{L}^{IP}(\boldsymbol{\lambda}^{\star}) = \vartheta(IP^{\star})$.*

5. *If $(\overline{IP})$ is feasible and $(\mathcal{L}_{\boldsymbol{\lambda}}^{IP})$ verifies the integrality property then $(IP^{\star})$ is feasible and*

$$\vartheta(\overline{IP}) = \vartheta(\mathcal{L}_{\overline{\boldsymbol{\lambda}}}^{IP}) = \vartheta(\mathcal{DL}^{IP}) = \mathcal{L}^{IP}(\boldsymbol{\lambda}^{\star}) = \vartheta(IP^{\star}).$$

From the above theorem it is clear that the Lagrangian dual will provide a bound which is at least as good as the one provided by the linear relaxation. However, the bound given by the Lagrangian dual will not improve the one obtained by solving $(IP^{\star})$. From the above theorem it also seems doubtful why one should solve the Lagrangian dual whenever its Lagrangian relaxation satisfies the integrality property. Nevertheless, in many cases it is computationally more efficient to solve the Lagrangian dual than to use the standard linear programming methods to solve the linear relaxation.

An important question that remains to be analyzed is how the Lagrangian dual $(\mathcal{DL}^{IP})$ can be optimized. Observe first that $\mathcal{L}^{IP}$ is the maximum of a finite number of linear functions. Hence, $\mathcal{L}^{IP}$ is given by the upper envelope of this finite set of linear functions, and therefore it is a piecewise linear convex function of $\boldsymbol{\lambda}$. This property is very important since it guarantees that any local minimum is also global. Moreover, this function is continuous on its domain and differentiable nearly everywhere, except at its breaking points. In fact, $\mathcal{L}^{IP}(\boldsymbol{\lambda})$ is differentiable at $\boldsymbol{\lambda}$ whenever the optimal solution set of $(\mathcal{L}_{\boldsymbol{\lambda}}^{IP})$ is a singleton. In this case its gradient corresponds to $\boldsymbol{b} - \boldsymbol{Ax}_{\boldsymbol{\lambda}}$ with $\boldsymbol{x}_{\boldsymbol{\lambda}}$ the unique optimal solution to $(\mathcal{L}_{\boldsymbol{\lambda}}^{IP})$. On the other hand, at the nondifferentiable points of $\mathcal{L}^{IP}(\boldsymbol{\lambda})$, i.e. points where the optimal solution set of $(\mathcal{L}_{\boldsymbol{\lambda}}^{IP})$ is not a singleton, the function is subdifferentiable. A subgradient of the function at $\boldsymbol{\lambda}$ is given by $\boldsymbol{b} - \boldsymbol{Ax}_{\boldsymbol{\lambda}}$ with $\boldsymbol{x}_{\boldsymbol{\lambda}}$ an optimal solution to $(\mathcal{L}_{\boldsymbol{\lambda}}^{IP})$. These properties of the dual problem suggest immediately the application of a gradient-type method, where at the nondifferentiable points an arbitrary subgradient is used. This is the basic idea behind the subgradient method, one of the most popular methods used to optimize Lagrangian duals. The subgradient method, also known as the generalized gradient optimization method, was first proposed and actively studied by researchers in the former Soviet Union, like Poljak [107] and Shor [126]. Independently, the subgradient method was also discovered by Held and Karp [71]. An extensive survey on general methods to optimize nondifferentiable functions and in particular on the theory of subgradient methods and their convergence proofs can be found in Shor [127]. References more oriented to practical issues are for instance Fisher [50] and Held *et al.* [72]. The subgradient method is an iterative method which starting with an initial λ_0, solves the associated Lagrangian relaxation and determines the next iteration point using the updating formula

$$\lambda_{k+1} := \max\left\{0, \lambda_k - t_k\left(b - Ax_{\lambda_k}\right)\right\} \tag{2.35}$$

where t_k is a scalar stepsize and x_{λ_k} an arbitrary optimal solution to $(\mathcal{L}_\lambda^{IP})$. The behavior of this method depends mostly on the way the parameter t_k is chosen. In fact, the standard results establishing the convergence of the subgradient method demand that, see Poljak [107]

$$t_k > 0, \lim_{k\uparrow\infty} t_k = 0 \text{ and } \sum_{k=0}^{\infty} t_k = +\infty.$$

Held *et al.* [72] proposed the following formula for t_k

$$t_k := \frac{p_k(\mathcal{L}^{IP}(\lambda_k) - \vartheta(IP))}{\|b - Ax_{\lambda_k}\|^2}$$

with $0 < p_k < 2$. The convergence of the subgradient method is still maintained if $\vartheta(IP)$ is replaced by a valid lower bound in the above formula. The sequence of p_k is usually initiated by taking $p_0 = 2$ and reducing it to half whenever $\mathcal{L}^{IP}(\lambda_k)$ fails to decrease in a specified number of iterations. This rule performs well in practice, although it does not satisfy the above sufficient condition to achieve convergence. An important issue in the subgradient method is the stopping rule. Actually, there is no easy way of proving optimality in the subgradient method. Observe that such a method would involve checking if the null vector belonged to the subgradient set at the current iteration point. Although it is possible to characterize completely the subgradient set at λ by

$$\text{co}\left\{b - Ax_\lambda : x_\lambda \text{ is an optimal solution to } (\mathcal{L}_\lambda^{IP})\right\}$$

checking if the null vector belongs to this set is rather difficult. In practice, the method is stopped after performing a specified number of iterations.

When the Lagrangian relaxation satisfies the integrality property, it follows from Theorem 2.4.1 that the Lagrangian dual will not provide a better bound than the linear relaxation. However, it may still be worthwhile to optimize this Lagrangian dual instead of solving directly the linear relaxation whenever the linear relaxation is highly degenerated. In particular, the robustness of the subgradient method applied to this Lagrangian dual can be improved. Observe by Theorem 2.4.1 that the corresponding variables in the linear dual are optimal multipliers to the Lagrangian dual. Therefore, the knowledge of the feasible set of the linear dual can be used to derive individual bounds on each Lagrangian multiplier. Let this bounding region be given by $\mathcal{B} \subseteq I\!\!R_+^m$ where $I\!\!R_+^m$ denotes the nonnegative halfspace of $I\!\!R^m$ and $\mathcal{B} := \{\lambda \in I\!\!R_+^m : l \leq \lambda \leq u\}$. Clearly, minimizing $(\mathcal{L}_\lambda^{IP})$ in $\mathcal{B}$ is equivalent to minimizing it on $I\!\!R_+^m$. Hence, we can solve $(\mathcal{D}\mathcal{L}^{IP})$ by minimizing the following modified Lagrangian function $\mathcal{L}_\mathcal{B}^{IP} : I\!\!R_+^m \longrightarrow I\!\!R \cup \{+\infty\}$ given by

$$\mathcal{L}_\mathcal{B}^{IP}(\lambda) = \begin{cases} \mathcal{L}^{IP}(\lambda) & \text{if } \lambda \in \mathcal{B} \\ +\infty & \text{otherwise} \end{cases}$$

This function is still convex, continuous and nondifferentiable in $\mathcal{B}$. Moreover, the subgradient method can be applied to minimize the above Lagrangian function since the convergence of the subgradient method does not require finiteness assumptions, see Poljak [107]. Observe that since the domain of $\mathcal{L}_\mathcal{B}^{IP}$ is given by $\mathcal{B}$, the subgradient set is empty at $\lambda \notin \mathcal{B}$. Hence, the subgradient method will only consider iteration points belonging to $\mathcal{B}$, and thus it is essential to find a subgradient at any point $\lambda \in \mathcal{B}$. Denoting by $\partial(\mathcal{L}^{IP})(\lambda)$ the subgradient set of $\mathcal{L}^{IP}$ at λ, the following lemma provides the solution to this problem.

Lemma 2.4.1

For $\lambda_0 \in \mathcal{B}$, we have $\partial(\mathcal{L}^{IP})(\lambda_0) \subseteq \partial(\mathcal{L}_\mathcal{B}^{IP})(\lambda_0)$. Moreover, if λ_0 belongs to the boundary of $\mathcal{B}$ then $d_\mathcal{B}$ is also a subgradient of $\mathcal{L}_\mathcal{B}^{IP}$ at λ_0, with

$$d_{\mathcal{B}_i} := \begin{cases} d_i & \textit{if } i \notin I_l(\lambda_0) \textit{ and } i \notin I_u(\lambda_0) \\ 0 & \textit{otherwise} \end{cases},$$

$d \in \partial(\mathcal{L}^{IP})(\lambda_0)$ and $I_l(\lambda_0) := \{i : \lambda_{0_i} = l_i \textit{ and } d_i > 0\}$, $I_u(\lambda_0) := \{i : \lambda_{0_i} = u_i \textit{ and } d_i < 0\}$.

Proof: From the definition of $\mathcal{L}_{\mathcal{B}}^{IP}$ it follows for $\lambda_0 \in \mathcal{B}$ and $d \in \partial(\mathcal{L}^{IP})(\lambda_0)$ that

$$\mathcal{L}_{\mathcal{B}}^{IP}(\lambda) \geq \mathcal{L}^{IP}(\lambda) \geq \mathcal{L}^{IP}(\lambda_0) + d^\top(\lambda - \lambda_0) = \mathcal{L}_{\mathcal{B}}^{IP}(\lambda_0) + d^\top(\lambda - \lambda_0), \quad (2.36)$$

for any λ, Hence, d is also a subgradient of $\mathcal{L}_{\mathcal{B}}^{IP}$ at λ_0.

It is left to prove that $d_\mathcal{B}$ is a subgradient of $\mathcal{L}_{\mathcal{B}}^{IP}$ at λ_0 in the boundary of $\mathcal{B}$. Clearly, for $d_\mathcal{B}$ the subgradient inequality (2.36) holds for $\lambda \notin \mathcal{B}$. Considering now the case that $\lambda \in \mathcal{B}$, observe that for $i \in I_u(\lambda_0)$ we have that $d_i(\lambda_i - u_i) \geq 0 = d_{\mathcal{B}_i}(\lambda_i - \lambda_{0_i})$, while for $i \in I_l(\lambda_0)$ we have that $d_i(\lambda_i - l_i) \geq 0 = d_{\mathcal{B}_i}(\lambda_i - \lambda_{0_i})$. Hence, $d^\top(\lambda - \lambda_0) \geq d_\mathcal{B}^\top(\lambda - \lambda_0)$, for all $\lambda \in \mathcal{B}$, and thus:

$$\mathcal{L}_{\mathcal{B}}^{IP}(\lambda) \geq \mathcal{L}_{\mathcal{B}}^{IP}(\lambda_0) + d_\mathcal{B}^\top(\lambda - \lambda_0),$$

which concludes the proof. $\qquad\qquad\square$

The above lemma induces the following changes in the multipliers updating formula (2.35)

$$\lambda_{k+1} := \min\left\{u, \max\left\{l, \lambda_k - t_k\left(b - Ax_{\lambda_k}\right)\right\}\right\}. \quad (2.37)$$

It is reasonable to expect that this simple modification in the subgradient method, enforcing the multipliers to remain in the "interesting" region, will attenuate the zig-zag effect, and accelerate the convergence. This is confirmed by our computational results, see Section 2.4.7.

In recent years several modifications of the basic subgradient method were proposed. Among the first, Lemaréchal [88] and Wolfe [134] proposed the conjugate subgradient method. This method is a descent algorithm which seeks for an appropriate search direction upon the information provided by the subgradients calculated at a neighborhood ε of the current point. After finding the descent direction, a line search is performed to compute the next iteration point. However, this method which is also known as ε-descent, does not appear to be efficient in practice, see [74]. Following the same reasoning, bundle methods seem to be more efficient for real life applications. For these methods the approximation of the subgradient set is constructed with a collection of weighted subgradients obtained in previous iterations: the bundle. A search direction is now computed in this bundle and the line search along this direction aims to compute the next iteration point and also to enrich the bundle of subgradients. An important issue of bundle methods is the fact that they are computationally more expensive. More information has to be kept and an extra optimization problem, finding the vector with the smallest Euclidean norm, and computing a line search, has to be solved per iteration. Variations of this type of methods can be found, for instance, in Kiwiel [83] and Kim and Ahn [82]. Finally, it is important to mention that in the recent book of Hiriart-Urruty and Lemaréchal [74] different approaches to optimize nondifferentiable functions, in particular ε-descent and bundle methods, are exhaustively discussed.

Another type of approach to tackle $(\mathcal{DL}^{IP})$ is given by the so-called multiplier adjustment methods. These methods differ from the subgradient type methods in the following points. Firstly, they ensure a monotone decreasing sequence of upper bounds. Secondly, they only adjust a limited number of multipliers at each iteration, providing at the end a heuristic solution. The process of generating these upper bounds and adjusting the multipliers depends greatly on the dual problem, since it explores its structure and particularities. These characteristics explain why in general multiplier adjustment methods are faster than the subgradient method and why the bounds provided are not as good. However, depending on the applications, this type of approach may be worthwhile as illustrated by Erlenkotter's algorithm [48] for the uncapacitated facility location problem. On the whole, subgradient optimization is an easy and robust technique based on nonlinear programming theory which usually works well, while the success of a multiplier adjustment method is very dependent on the problem itself.

Finally, it is important to mention that general techniques like the simplex method with column generation, the ellipsoid method, cutting plane algorithms and interior point algorithms can also be applied to solving the dual problem. These methods are in general difficult to implement and require per iteration more computations than the simpler subgradient method.

In the next subsections we will consider for each of the formulations proposed in Section 2.4.2, a specific Lagrangian relaxation. These Lagrangian duals can be solved using the described modified subgradient method.

Lagrangian Relaxation of (SF)

Recalling formulation (SF), it appears that the "complicating" constraints are (2.14), (2.21) and (2.22). Hence, we will relax these blocks of constraints and consider the associated set of multipliers $\boldsymbol{\nu} := (\boldsymbol{\lambda}, \boldsymbol{\alpha}, \boldsymbol{\beta})$ where λ_i are free variables for every $i \in I$, $\alpha_{ij} \geq 0$, $i \in I, j \in J$ and $\beta_{ik} \geq 0$, $i \in I, k \in K$.

Notice that since the only elements of the constraint matrix of each subproblem are 0 and ± 1, and since in each constraint (row of the matrix) there are only two nonzero elements, 1 and -1, by one of the sufficient conditions listed in [102] this matrix is totally unimodular. Hence, from Theorem 2.4.1 it follows that this Lagrangian relaxation satisfies the integrality property, and thus provides an optimal value equal to the linear relaxation $(\overline{SF})$ bound. In spite of the fact that these bounds coincide, it is computationally more efficient to solve this Lagrangian dual, since each subproblem corresponding to a given set of multipliers can be solved by inspection of the objective function. In fact, for each set of multipliers the Lagrangian relaxation is given by

$$
\mathcal{L}^{SF}(\boldsymbol{\nu}) = \sum_{i \in I} \lambda_i +
$$

$$
\max \quad \sum_{i \in I} \sum_{j \in J} \sum_{k \in K} (c_{ijk} - \lambda_i - \alpha_{ij} - \beta_{ik}) x_{ijk} - \sum_{j \in J} \sum_{k \in K} F_{jk} t_{jk} +
$$

$$
\sum_{j \in J} \left(\sum_{i \in I} \alpha_{ij} - f_j \right) y_j + \sum_{k \in K} \left(\sum_{i \in I} \beta_{ik} - g_k \right) z_k
$$

$$
\text{s.t.:} \quad (2.15), (2.18), (2.19), (2.20).
$$

From (2.15), (2.18), (2.20) and the nature of the objective function we can simplify $\mathcal{L}^{SF}(\boldsymbol{\nu})$. Introducing the *reduced costs* $\hat{c}_{jk} := \sum_{i \in I}(c_{ijk} - \lambda_i - \alpha_{ij} - \beta_{ik})^+$, $\hat{f}_j := (\sum_{i \in I} \alpha_{ij} - f_j)^+$ and $\hat{g}_k := (\sum_{i \in I} \beta_{ik} - g_k)^+$, where $(x)^+ := \max\{0, x\}$, we get

$$
\mathcal{L}^{SF}(\boldsymbol{\nu}) = \sum_{i \in I} \lambda_i + \sum_{j \in J} \hat{f}_j + \sum_{k \in K} \hat{g}_k + \max \sum_{j \in J} \sum_{k \in K} (\hat{c}_{jk} - F_{jk}) t_{jk}
$$

$$
\text{s.t.:} \quad (2.19).
$$

Using now the fact that the t_{jk} are binary variables and taking into account the nature of the objective function, we finally obtain the following analytical expression for $\mathcal{L}^{SF}(\nu)$

$$\mathcal{L}^{SF}(\nu) = \sum_{i \in I} \lambda_i + \sum_{j \in J} \hat{f}_j + \sum_{k \in K} \hat{g}_k + \sum_{j \in J} \sum_{k \in K} (\hat{c}_{jk} - F_{jk})^+. \tag{2.38}$$

The optimal values of the variables are given by

$$y_j = \begin{cases} 1 & \text{if } \hat{f}_j > 0 \\ 0 & \text{otherwise} \end{cases},$$

$$z_k = \begin{cases} 1 & \text{if } \hat{g}_k > 0 \\ 0 & \text{otherwise} \end{cases},$$

$$t_{jk} = \begin{cases} 1 & \text{if } \hat{c}_{jk} - F_{jk} > 0 \\ 0 & \text{otherwise} \end{cases},$$

$$x_{ijk} = \begin{cases} 1 & t_{jk} = 1 \text{ and } c_{ijk} - \lambda_i - \alpha_{ij} - \beta_{ik} > 0 \\ 0 & \text{otherwise} \end{cases}$$

It is interesting to observe that all the simplifications made in $\mathcal{L}^{SF}(\nu)$ can also be justified via the dual problem of $(\overline{SF})$. Notice that solving each $(\mathcal{L}^{SF}_\nu)$ has the same complexity order as the one required to update the multipliers, i.e. $\mathcal{O}(m(p+q))$.

In order to solve the Lagrangian dual $(\mathcal{DL}^{SF})$ we consider the modified subgradient method described previously. Therefore, at each iteration of the subgradient method the updating formula is modified so that the Lagrangian multipliers remain inside the closed region defined by (2.29), (2.33) and (2.34).

This approach can also be used, with the correspondent simplifications, in both special cases, the 2-level uncapacitated facility location problem and the 2-echelon uncapacitated facility location problem. For the 2-echelon uncapacitated facility location problem the Lagrangian relaxation can also be solved by inspection and its optimal value is given by

$$\sum_{i \in I} \lambda_i + \sum_{j \in J} \sum_{k \in K} \left(\sum_{i \in I} (c_{ijk} - \lambda_i - \alpha_{ij})^+ - F_{jk} \right)^+ + \sum_{j \in J} \left(\sum_{i \in I} \alpha_{ij} - f_j \right)^+.$$

Also for the 2-level uncapacitated facility location problem solving the Lagrangian relaxation can be done by inspection and its optimal value is given by

$$\sum_{i \in I} \lambda_i + \sum_{j \in J} \sum_{k \in K} \sum_{i \in I} (c_{ijk} - \lambda_i - \alpha_{ij} - \beta_{ik})^+ +$$

$$\sum_{j \in J} \left(\sum_{i \in I} \alpha_{ij} - f_j \right)^+ + \sum_{k \in K} \left(\sum_{i \in I} \beta_{ik} - g_k \right)^+ .$$

By solving the duals corresponding to these Lagrangian relaxations we will obtain bounds which are never worse than the ones previously reported in [59, 60, 131]. In fact, the bounds derived in these papers are obtained from the corresponding (WF).

Lagrangian Relaxation of (BSF)

Although $(\mathcal{DL}^{SF})$ can easily be solved by the subgradient method (each Lagrangian relaxation has an analytical expression for the optimal solution), the domain of each subproblem is very unrestricted. Consequently, the number of feasible solutions of each subproblem is substantial as well as the number of alternative optimal solutions. Observe that alternative optimal solutions occur whenever $x = 0$ for at least one $(x)^+$. This means that the probability of finding a nondifferentiability point of $\mathcal{L}^{SF}$ is high and that the size of the subgradient set, the subdifferential, is big, see [104]. These characteristics may contribute to practical difficulties in the convergence of the subgradient method. Therefore, we will consider now the Lagrangian relaxation of (BSF) where constraints (2.14), (2.21) and (2.22) are relaxed. Again, the integrality property holds since the constraint matrix of $(\mathcal{L}_{\nu}^{BSF})$ remains totally unimodular, and therefore by solving this Lagrangian dual $(\mathcal{DL}^{BSF})$ we will not obtain a better bound than $\vartheta(\overline{BSF})$. Nevertheless, it is interesting to analyze if this Lagrangian relaxation will be as easy to solve as $(\mathcal{L}_{\nu}^{SF})$. Now, for each set of multipliers $\nu = (\lambda, \alpha, \beta)$, where λ_i are again free variables for $i \in I$, $\alpha_{ij} \geq 0$, $i \in I, J \in J$ and $\beta_{ik} \geq 0$, $i \in I, j \in J, k \in K$, we will see that the corresponding $(\mathcal{L}_{\nu}^{BSF})$

$$\mathcal{L}^{BSF}(\nu) = \sum_{i \in I} \lambda_i +$$

$$\max \sum_{i \in I} \sum_{j \in J} \sum_{k \in K} (c_{ijk} - \lambda_i) x_{ijk} - \sum_{j \in J} \sum_{k \in K} F_{jk} t_{jk} - \sum_{j \in J} f_j y_j - \sum_{k \in K} g_k z_k +$$

$$\sum_{i \in I} \sum_{j \in J} \alpha_{ij} \left(y_j - \sum_{k \in K} x_{ijk} \right) + \sum_{i \in I} \sum_{k \in K} \beta_{ik} \left(z_k - \sum_{j \in J} x_{ijk} \right)$$

$$= \sum_{i \in I} \lambda_i + \max \sum_{i \in I} \sum_{j \in J} \sum_{k \in K} (c_{ijk} - \lambda_i - \alpha_{ij} - \beta_{ik}) x_{ijk} - \sum_{j \in J} \sum_{k \in K} F_{jk} t_{jk} -$$

$$\sum_{j \in J} \left(f_j - \sum_{i \in I} \alpha_{ij} \right) y_j - \sum_{k \in K} \left(g_k - \sum_{i \in I} \beta_{ik} \right) z_k$$

s.t.: $(2.15), (2.16), (2.17), (2.18), (2.19), (2.20)$

is equivalent to a *min-cut problem*.

Theorem 2.4.2

For each set of multipliers ν, the corresponding Lagrangian relaxation $(\mathcal{L}_{\nu}^{BSF})$ is equivalent to a min-cut problem.

Proof: In order to prove this result some rewriting of the problem is needed. Due to the nature of the objective function, some variables will be fixed to 1 by examining the following *reduced costs*

(i) If $f_j - \sum_{i \in I} \alpha_{ij} < 0$ then in any optimal solution, facility j will be open;

(ii) If $g_k - \sum_{i \in I} \beta_{ik} < 0$ then in any optimal solution, depot k will be open.

Fixing a facility j (depot k) to be open is achieved by replacing its "reduced" cost by zero and subtracting the original negative value from the value of the objective function after solving the subproblem. We will henceforth assume that these simplifications have been made, i.e. all the "reduced" costs of y_j and z_k are nonnegative.

By (2.15), (2.19) and (2.20) and the nature of the objective function we can set

$$x_{ijk} = \begin{cases} t_{jk} & \text{if } c_{ijk} - \lambda_i - \alpha_{ij} - \beta_{ik} > 0 \\ 0 & \text{otherwise} \end{cases}.$$

Hence, we obtain

$$\mathcal{L}^{BSF}(\nu) = \sum_{i \in I} \lambda_i + \max \sum_{j \in J} \sum_{k \in K} \sum_{i \in I} (c_{ijk} - \lambda_i - \alpha_{ij} - \beta_{ik})^+ t_{jk} - \sum_{j \in J} \sum_{k \in K} F_{jk} t_{jk} -$$

$$\sum_{j \in J} \left(f_j - \sum_{i \in I} \alpha_{ij} \right) y_j - \sum_{k \in K} \left(g_k - \sum_{i \in I} \beta_{ik} \right) z_k$$

s.t.: $(2.16), (2.17), (2.18), (2.19)$.

Considering again $\hat{c}_{jk} := \sum_{i \in I}(c_{ijk} - \lambda_i - \alpha_{ij} - \beta_{ik})^+$ and defining $f'_j := f_j - \sum_{i \in I} \alpha_{ij}$ and $g'_k := g_k - \sum_{i \in I} \beta_{ik}$, we can rewrite this subproblem as follows

$$\mathcal{L}^{BSF}(\nu) = \sum_{i \in I} \lambda_i + \max \sum_{j \in J}\sum_{k \in K}(\hat{c}_{jk} - F_{jk})t_{jk} - \sum_{j \in J} f'_j y_j - \sum_{k \in K} g'_k z_k$$
$$\text{s.t.: } (2.16),(2.17),(2.18),(2.19).$$

Notice that if $(\hat{c}_{jk} - F_{jk}) < 0$ then it is not profitable to have the pair (j,k) operating together, and so $t_{jk} = 0$ in any optimal solution to $(\mathcal{L}^{BSF}_\nu)$.

This subproblem can be further reduced by dropping some other variables

(iii) If $\sum_{k \in K}(\hat{c}_{jk} - F_{jk})^+ < f'_j$ then opening facility j is not profitable, so $y_j = 0$ in any optimal solution to $(\mathcal{L}^{BSF}_\nu)$;

(iv) If $\sum_{j \in J}(\hat{c}_{jk} - F_{jk})^+ < g'_k$ then opening depot k is not profitable, so $z_k = 0$ in any optimal solution to $(\mathcal{L}^{BSF}_\nu)$.

We can cycle through these reductions and update the set of possible sites until no further simplifications can be made. Observe that a variable is fixed to zero by simply removing it from the problem.

Setting now $c'_{jk} := (\hat{c}_{jk} - F_{jk})^+$ we can rewrite $\mathcal{L}^{BSF}(\nu)$ as

$$\mathcal{L}^{BSF}(\nu) = \sum_{i \in I} \lambda_i + \max \sum_{j \in J}\sum_{k \in K} c'_{jk} t_{jk} - \sum_{j \in J} f'_j y_j - \sum_{k \in K} g'_k z_k$$
$$\text{s.t.: } (2.16),(2.17),(2.18),(2.19).$$

By defining $f''_j := \sum_{k \in K} c'_{jk} - f'_j$ (by reduction (iii) this value is nonnegative), using $\max f = -\min -f$, and by rearranging the terms in the objective function, we obtain

$$\mathcal{L}^{BSF}(\nu) = \sum_{i \in I} \lambda_i - \min \sum_{j \in J}\sum_{k \in K} -c'_{jk}t_{jk} + \sum_{j \in J}\left(\sum_{k \in K} c'_{jk} - f''_j\right) y_j + \sum_{k \in K} g'_k z_k$$
$$= \sum_{i \in I} \lambda_i - \min \sum_{j \in J}\sum_{k \in K} c'_{jk}(y_j - t_{jk}) - \sum_{j \in J} f''_j y_j + \sum_{k \in K} g'_k z_k$$
$$= \sum_{i \in I} \lambda_i + \sum_{j \in J} f''_j -$$

$$\min \sum_{j \in J} \sum_{k \in K} c'_{jk}(y_j - t_{jk}) + \sum_{j \in J} f''_j(1 - y_j) + \sum_{k \in K} g'_k z_k$$

$$\text{s.t.: } (2.16), (2.17), (2.18), (2.19).$$

Introducing the variables y_s and z_t, defining

$$w_{sj} = 1 - y_j \geq 0 \quad \text{from} \quad (2.18) \qquad\qquad (2.39)$$

$$w_{jk} = y_j - t_{jk} \geq 0 \quad \text{from} \quad (2.16) \qquad\qquad (2.40)$$

$$w_{kt} = \quad z_k \quad \geq 0 \quad \text{from} \quad (2.18) \qquad\qquad (2.41)$$

and using (2.17) this Lagrangian relaxation can be rewritten as

$$\mathcal{L}^{BSF}(\nu) = \sum_{i \in I} \lambda_i + \sum_{j \in J} f''_j -$$

$$\min \quad \left\{ \sum_{j \in J} f''_j w_{sj} + \sum_{j \in J} \sum_{k \in K} c'_{jk} w_{jk} + \sum_{k \in K} g'_k w_{kt} \right\}$$

$$\text{s.t.: } \quad y_s - z_t \geq 1$$

$$z_k - t_{jk} = -y_j + w_{jk} + z_k \geq 0 \quad \forall j \in J, k \in K$$

$$-y_s + w_{sj} + y_j \geq 0 \quad \forall j \in J$$

$$-z_k + w_{kt} + z_t \geq 0 \quad \forall k \in K$$

$$w_{sj}, w_{jk}, w_{kt} \geq 0 \quad \forall j \in J, k \in K$$

which is a min-cut problem on the graph given in Figure 2.3, with s the source, t the sink, and $w_{ab} = 1$ if arc (a, b) belongs to the cut. $\qquad\qquad\square$

Hence, using the above theorem, for each set of multipliers ν we simply need to solve a min-cut problem. This min-cut problem arises in a very special network, which will be denoted by network$_f$, see Figure 2.3.

From the optimal $s - t$ cut-set, $(S, \overline{S})$ and using (2.39), (2.40) and (2.41), an optimal solution to $(\mathcal{L}_\nu^{BSF})$ can be constructed in the following way,

$$y_j = \begin{cases} 1 & \text{if } j \in S \\ 0 & \text{if } j \in \overline{S} \end{cases},$$

$$z_k = \begin{cases} 1 & \text{if } k \in S \\ 0 & \text{if } k \in \overline{S} \end{cases},$$

$$t_{jk} = \begin{cases} 1 & \text{if } y_j = z_k = 1 \text{ and } c'_{jk} > 0 \\ 0 & \text{otherwise} \end{cases},$$

$$x_{ijk} = \begin{cases} t_{jk} & \text{if } c_{ijk} - \lambda_i - \alpha_{ij} - \beta_{ik} > 0 \\ 0 & \text{otherwise} \end{cases}.$$

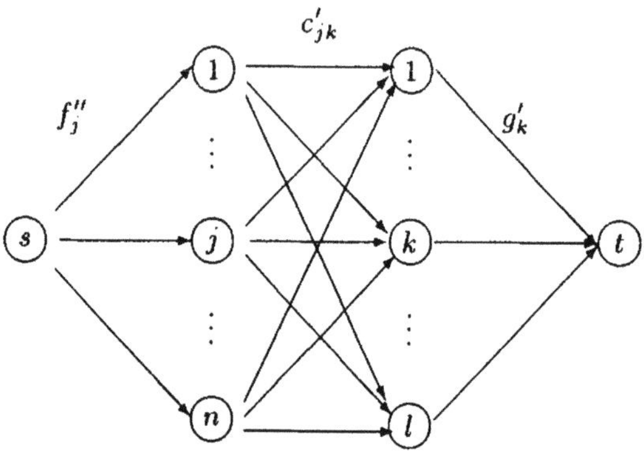

Figure 2.3: Network$_f$.

Observe that by following the same reasoning and setting $g''_k := \sum_{j \in J} c'_{jk} - g'_k$ (due to (iv) this value is also nonnegative), we can establish, for each set of multipliers, another equivalent min-cut problem in a similar network, say network$_g$.

In order to solve the Lagrangian dual $(\mathcal{DL}^{BSF})$, we use again the modified subgradient method by limiting the variation of the Lagrangian multipliers to the bounds derived via the dual of $(\overline{BSF})$: (2.29), (2.33) and (2.34).

Notice that the effort required to solve each $(\mathcal{L}_\nu^{BSF})$ corresponds to updating the multipliers, $\mathcal{O}(m(p+q))$, and solving the min-cut problem. This problem is the dual of the *max-flow problem* for which several algorithms are available, see [2]. Phillips and Dessouky [105] proposed a method to locate directly the optimal cut set in a network with lower bounds on the arc capacities. Basically, this algorithm starts by having on the source side of the cut all the nodes except the sink t, and then tries to decide which nodes do not belong to the source side of the optimal cut. Due to the structure of the network$_f$ and relations $\sum_{j \in J} c'_{jk} \geq g'_k$ and $\sum_{k \in K} c'_{jk} \geq f''_j$, this algorithm can easily be adapted. However, its running time cannot be guaranteed to be polynomial. Among the available algorithms for solving the max-flow problem the preflow-push algorithms are widely used due to their simplicity and flexibility, see [2]. This type of algorithms seems to be particularly adequate to our problem since they first detect the minimum cut and only in a second phase obtain the max-flow. Moreover, their running time depends mostly on the way certain steps are performed. Among them the highest-label preflow-push algorithm with a running time of $\mathcal{O}(n^2\sqrt{m})$, with n the total number of nodes and m the number of arcs, seems to perform quite well in practice. Another important point, is that during the subgradient optimization, some "easy" forms of the network$_f$ can appear for which the minimum cut is trivially found. Observe that for each set of multipliers, the resulting network$_f$ varies and thus, we cannot predict in advance if one of the node sets will be bigger than the other. Therefore, although the min-cut problem to be solved at each iteration is defined on a bipartide network, a specialized bipartite preflow-push algorithm may not be more efficient than a preflow-push algorithm to solve this problem, see [2].

In spite of requiring more computational time per iteration, during the subgradient optimization, it is reasonable to expect a smaller number of iterations to obtain the same bounds than when solving $(\mathcal{DL}^{SF})$. This is supported by the fact that since this Lagrangian relaxation possesses more structure, less alternative optimal solutions occur, and therefore we may expect a better performance of the subgradient method applied to this case.

The results derived above are still valid, with the correspondent simplifications, in both special cases, the 2-level uncapacitated facility location problem and the 2-echelon uncapacitated facility location problem. For the 2-echelon uncapacitated facility location problem, the corresponding Lagrangian relaxation can now be solved by inspection and its optimal value is given by

$$\sum_{i\in I}\lambda_i + \sum_{j\in J}\left(\sum_{k\in K}\left(\sum_{i\in I}(c_{ijk}-\lambda_i-\alpha_{ij})^+ - F_{jk}\right)^+ + \left(\sum_{i\in I}\alpha_{ij}-f_j\right)\right)^+ .$$

For the 2-level uncapacitated facility location problem, the corresponding Lagrangian relaxation is also equivalent to a min-cut problem. This result is easily proved by maintaining in the formulation the variables t_{jk} with $F_{jk} = 0$ and following the scheme of the proof of Theorem 2.4.2. Hence, the network has the same structure as the network$_f$ with different arc capacities: the arcs connecting facilities and depots will have capacities given by $\hat{c}_{jk}$, the arcs connecting the source node to the facilities have capacities given by $\sum_{k \in K} \hat{c}_{jk} - f'_j$ and finally the arcs connecting the depots to the sink node have capacities given by g'_k.

Again, by solving these duals we will obtain bounds that will never be worse than the ones previously reported in [59, 60, 131].

Lagrangian Relaxation of (WF)

Although Proposition 2.4.1 indicates that formulation (WF) in terms of the linear bounds will never be better than (SF) and (BSF), it is nevertheless interesting to study its potential in terms of Lagrangian relaxationt. Dualizing the block (2.14) of constraints in formulation (WF) we obtain the following Lagrangian relaxation

$$\mathcal{L}^{WF}(\lambda) = \sum_{i \in I} \lambda_i + \max \sum_{i \in I} \sum_{j \in J} \sum_{k \in K} (c_{ijk} - \lambda_i) x_{ijk} -$$

$$\sum_{j \in J} f_j y_j - \sum_{k \in K} g_k z_k - \sum_{j \in J} \sum_{k \in K} F_{jk} t_{jk}$$

$$\text{s.t.: } (2.15), (2.16), (2.17), (2.18), (2.19), (2.20).$$

Since the matrix formed by constraints (2.15), (2.16) and (2.17) is totally unimodular, this Lagrangian relaxation also satisfies the integrality property. Hence, from Theorem 2.4.1 it follows that the optimal value of $(\mathcal{DL}^{WF})$ will be equal to $\vartheta(\overline{WF})$.

Though the bounds provided by $(\overline{WF})$ will not be better than the ones obtained by solving $(\overline{SF})$ or $(\overline{BSF})$, it is important to test in practice if there exists a significant difference in quality. Therefore, we will also analyze the Lagrangian dual $(\mathcal{DL}^{WF})$. Using a similar reasoning as described in the proof of Theorem 2.4.2, it follows that the above Lagrangian relaxation is equivalent to $(\mathcal{L}^{BSF}_\nu)$ with all α_{ij} and β_{ik} equal to zero, and thus it is also a min-cut problem. The corresponding network has the same structure as the network$_f$ with arc capacities for the arcs connecting facilities and depots given by $\left(\sum_{i \in I}(c_{ijk} - \lambda_i)^+ - F_{jk}\right)^+$, while the capacities on the arcs connecting the source node to the facilities are given by $\sum_{k \in K} \left(\sum_{i \in I}(c_{ijk} - \lambda_i)^+ - F_{jk}\right)^+ - f_j$. Finally, the arcs connecting the depots to the sink node have capacities g_k. Observe that constructing this network is less time consuming than in the previous case since there is only one set of multipliers, $\boldsymbol{\lambda}$.

In order to solve this Lagrangian dual $(\mathcal{DL}^{WF})$, we use again the modified subgradient method, by limiting the variation of the Lagrangian multipliers to the bounds derived via the dual of the linear relaxation (2.29). Observe that although a special min-cut problem has to be solved at each iteration of the subgradient method, the effort of updating the multipliers is only $\mathcal{O}(m)$.

In both special cases, the 2-level uncapacitated facility location problem and the 2-echelon uncapacitated facility location problem, the approaches found in the literature only approximate the "weak" linear relaxation of the corresponding (WF), by using a heuristic derived from the dual adjustment method of Erlenkotter [48]. Therefore, it is important to describe our approach in order to obtain the same bound. For the 2-echelon uncapacitated facility location problem, the correspondent Lagrangian relaxation can be solved by inspection and its optimal value is given by

$$\mathcal{L}^{WF}(\boldsymbol{\nu}) = \sum_{i \in I} \lambda_i + \sum_{j \in J} \left(\sum_{k \in K} \left(\sum_{i \in I} (c_{ijk} - \lambda_i)^+ - F_{jk} \right)^+ - f_j \right)^+ .$$

As for the 2-level uncapacitated facility location problem, the Lagrangian relaxation will also correspond to a min-cut problem. The network has the same structure as the above described network with the obvious simplification on the arcs capacities yielded by $F_{jk} = 0$.

2.4.5   Heuristics

For $\mathcal{NP}$-hard problems it is extremely important to find approximate solutions. An immediate way to obtain lower bounds for our problem is given by the extension of the greedy heuristic developed for the uncapacitated facility location problem in [86], in Algorithm 2.1.

This heuristic starts with all (facility,depot) pairs closed. Then, at each step, it selects from among all the possible pairs the one which, by having its components open and operating together, maximizes the increase in the objective function, i.e. the pair that yields the highest profit in that step. In order to compute these profits we need to take into account the fixed costs associated with opening entities (if one (both) of the components of the pair is (are) not yet open) and the fixed costs associated with having these entities operating together. When no further profit can be achieved by opening new pairs, an improvement step is applied. In this step we eliminate uneconomical pairs that were opened in earlier steps of the greedy heuristic.

Step 0. Let $J^\star = K^\star := \emptyset$ and $T^\star := \{(\emptyset, \emptyset)\}$;

Compute $\sigma_{j^\star k^\star} := \max_{(j,k) \in T} \sum_{i \in I} c_{ijk} - F_{jk} - f_j - g_k$;

Let $J^\star := \{j^\star\}; K^\star := \{k^\star\}; T^\star := \{(j^\star, k^\star)\}$;

For $i \in I$ **Do** $c_i^\star := c_{ij^\star k^\star}$;

Step 1. **For** $(j, k) \notin T^\star$ **Do**

    If $j \in J^\star$

        Then If $k \in K^\star$

            Then $\sigma_{jk} := \sum_{i \in I}(c_{ijk} - c_i^\star)^+ - F_{jk}$

            Else $\sigma_{jk} := \sum_{i \in I}(c_{ijk} - c_i^\star)^+ - F_{jk} - g_k$

        Else If $k \in K^\star$

            Then $\sigma_{jk} := \sum_{i \in I}(c_{ijk} - c_i^\star)^+ - F_{jk} - f_j$

            Else $\sigma_{jk} := \sum_{i \in I}(c_{ijk} - c_i^\star)^+ - F_{jk} - f_j - g_k$

    Compute $\sigma_{j^\star k^\star} := \max_{(j,k) \notin T^\star} \sigma_{jk}$;

Step 2. **If** $\sigma_{j^\star k^\star} > 0$

    Then Update $c_i^\star$;

        Let $T^\star := T^\star \cup \{(j^\star, k^\star)\}$;

        Let $J^\star := J^\star \cup \{j^\star\}$, $K^\star := K^\star \cup \{k^\star\}$ and **GoTo** *Step* 1

    Else GoTo *Step* 3;

Step 3. Consider the open (facility,depot) pairs by their order of opening; If the value of the solution is improved by closing a pair remove it from $T^\star$ until no more improvement can be achieved.

Algorithm 2.1: Greedy heuristic.

The complexity of Algorithm 2.1 is $\mathcal{O}(m\,(p\,q)^2)$, because the *Step* 1 is executed at most pq times with a complexity of $\mathcal{O}(m\,p\,q)$ in each iteration. The later is also the complexity of the initialization method as well as of the improvement phase in *Step* 3.

The performance of the greedy heuristic is not as good as the corresponding heuristic for the uncapacitated facility location problem, as our computational results show. This is explained by the fact that the objective function of the uncapacitated facility location problem is submodular, while in our problem this property is not satisfied. Therefore, it is important to devise other methods of deriving lower bounds. Since we proposed in the last section to solve a Lagrangian dual using a subgradient method, it seems logical to derive lower bound solutions via the Lagrangian relaxation. Although in practice the Lagrangian solutions are seldom feasible, they can be made feasible with some minor modifications. This technique has been successfully applied to several location problems as described by Beasley [20]. Therefore, we also considered lower bounds derived from the upper bound solutions obtained during the optimization of the Lagrangian duals described in Section 2.4.4. Given the (facility,depot) pairs set open in the Lagrangian solution, a feasible solution is immediately obtained by assigning each client to the most profitable (facility,depot) pair. This quite inexpensive method seems to perform well, as our computational results show.

Observe that for the special cases, the 2-level uncapacitated facility location problem and the 2-echelon uncapacitated facility location problem, the greddy heuristic described in Algorithm 2.1 is easily adapted to these problems. The same holds for the derivation of lower bound solutions from the corresponding Lagrangian solutions. Another approach to derive lower bounds to the 2-level uncapacitated facility location problem can be found in Tcha and Lee [131]. Their heuristic is a primal ascent method which tries to improve the primal solution given by the dual descent method. This approach was also followed by Gao and Robinson [59, 60] to derive lower bound solutions to the 2-echelon uncapacitated facility location problem.

2.4.6   Branch and Bound

A branch and bound algorithm is an enumerative method which mainly divides the problem into smaller problems and tries to solve these smaller problems by using bounding methods. Therefore, it is not only important to have good branching rules, i.e. choosing the variables to branch upon, but also good bounding methods. These bounding methods are particularly important since they permit to avoid certain non interesting nodes of the branch and bound tree. A node can be fathomed whenever the value of the optimal solution of the upper bounding method is smaller or equal than the value of the best lower bound solution. Nodes can also be fathomed when the feasible set of the problem to be solved is empty or when an optimal solution to the problem defined at the current node was found. Another important aspect is the use of penalties during the enumeration in order to remove "non-interesting" variables.

In this section we will present a branch an bound algorithm where the lower bounds are given by the best value achieved by the greedy-type heuristic or derived from the Lagrangian solutions as described in the last section. Upper bounds are obtained by solving $(\mathcal{DL}^{SF})$ presented in Section 2.4.4. Finally, in order to increase the efficiency of the branch and bound algorithm, we also consider penalty terms or penalties in order to close facilities and/or depots. We will exemplify the use of these penalties for the facilities. As we have seen in Section 2.4.4, for each set of multipliers ν the optimal solution of the Lagrangian relaxation $(\mathcal{L}_{\nu}^{SF})$ is such that a facility j is closed if its reduced cost is negative, i.e. $\hat{f}_j := \sum_{i \in I} \alpha_{ij} - f_j < 0$. Therefore, an upper bound on the optimal value of the Lagrangian dual where this facility j is open, is given by $\mathcal{L}^{SF}(\nu) + \hat{f}_j$. Consequently, whenever $\mathcal{L}^{SF}(\nu) + \hat{f}_j$ is smaller than the value of the best lower bound available, we can close this facility j. Similarly, penalties are derived for depots and operating pairs. Clearly, the efficiency of these penalties depends on how "close" the used multipliers are from the optimal ones associated with $(\mathcal{DL}^{SF})$.

Instead of using a binary branching rule by selecting one t_{jk} to branch, we developed a mixed branching rule that first considers a (facility,depot) pair, creating four mutually exclusive descendant nodes: $(y_j = 0$ and $z_k = 0)$, $(y_j = 0$ and $z_k = 1)$, $(y_j = 1$ and $z_k = 0)$ and $(y_j = 1$ and $z_k = 1)$, as shown in Figure 2.4. If at a branching node only variables of one type of entity remain free, by branching on one of these variables only one pair of descendant nodes is created. Notice also that at a node where $y_j = 0$ or $z_k = 0$, variables t_{jk} are fixed to zero respectively for all $k \in K$ or $j \in J$. When there are no more free facilities and depots, and the current node cannot be fathomed, we start branching on the remaining variables, i.e. those

t_{jk} for which both $y_j = 1$ and $z_k = 1$, see Figure 2.4. Notice that from this node on,

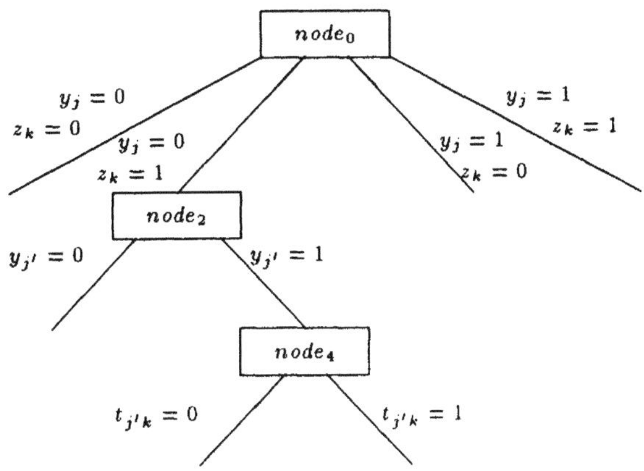

Figure 2.4: Branching rule.

the problem to be solved reduces to an instance of an uncapacitated facility location problem where the facilities to be located are the allowed pairs t_{jk}. Although the maximum number of nodes potentially created by this branching rule is bigger than that of the binary rule on only t_{jk}, our branching rule seems to perform better in practice. This may be due to the fact that the mixed rule induces in early stages a stronger partition in the feasible set of the problem than the binary rule.

At each node, and if no stopping rule applies, the variables to branch upon are chosen according to the biggest reduced cost as defined in Section 2.4.4. The purpose of this rule is to try to quickly decrease the upper bounds.

The branching is carried out in a depth-first search manner and was designed so that the optimization of each node takes advantage of the computational effort done at previous nodes, see Fisher [50].

In the special cases, the 2-level uncapacitated facility location problem and the 2-echelon uncapacitated facility location problem, the described branch and bound algorithm, can be simplified due to the specific characteristics of these problems. In both cases we embedded in the branch and bound algorithm the bounds obtained using the corresponding strong formulations (SF). However, the branching rules considered for these special cases are different. For the 2-level uncapacitated facility location problem we used a branching rule only on y_j and z_k variables, see Figure 2.5, since variables t_{jk} do not exist in this model. However, t_{jk} can be artificially intro-

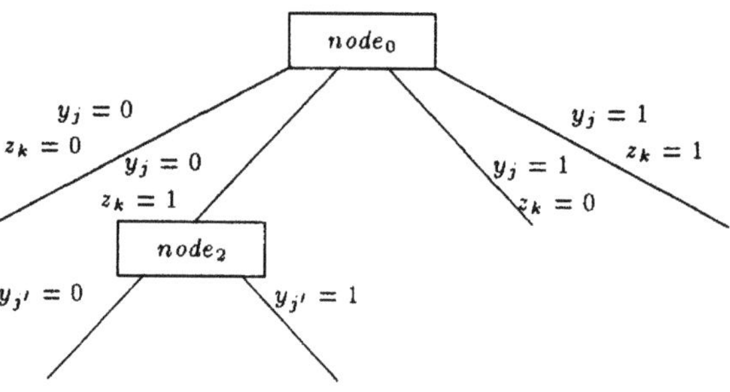

Figure 2.5: Branching rule for the 2-level case.

duced by letting t_{jk} represent the product of y_j and z_k. It appears that a binary rule on $t_{jk} = y_j z_k = 0$ or $t_{jk} = y_j z_k = 1$ is equivalent to the branching rule presented above, i.e. $t_{jk} = y_j z_k = 1$ corresponds to $y_j = z_k = 1$ while $t_{jk} = y_j z_k = 0$ corresponds to the partition $y_j = z_k = 0$, $y_j = 0$ and $z_k = 1$, $y_j = 1$ and $z_k = 0$. Tcha and Lee [131] embedded the primal ascent and dual descent heuristic, together with an incorrect node simplification rule, in a binary branch and bound algorithm, on variables y_j and z_k, see [18]. Observe also that the maximum number of nodes potentially created by our branching rule, is smaller for $p, q \geq 2$ than the corresponding number from the binary branching rule on y_j and z_k presented in [131].

Finally, for the 2-echelon uncapacitated facility location problem we developed a branching rule only on variables y_j and t_{jk}. Hence, branching is done on variables y_j until there are no more free facilities and the current node cannot be fathomed. Only then branching on (the remaining) t_{jk} is initiated, see Figure 2.6.

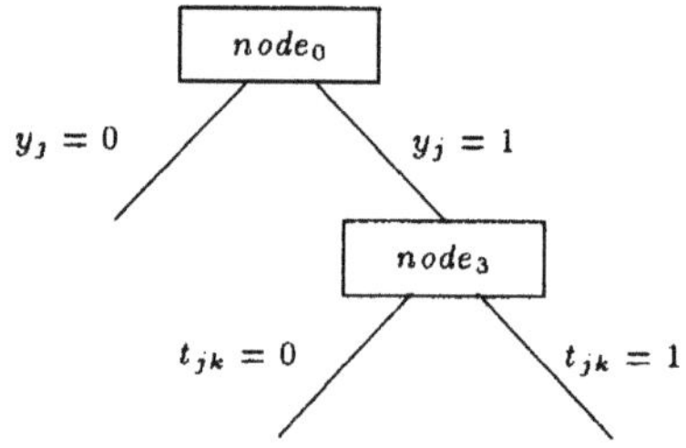

Figure 2.6: Branching rule for 2-echelon case.

However, in this case it is more efficient to use the simpler binary branching rule on t_{jk}, as it can be seen in our computational experience. Gao and Robinson [59, 60] report on how their extension of the dual descent and adjustment method was embedded in a binary branch and bound algorithm, on variables y_j and t_{jk}. For this particular case we can expect that the branch and bound algorithm described above will be more efficient since it uses stronger bounds.

2.4.7 Computational Results

The computational experience presented in this section was designed in order to test and compare the approaches discussed in the previous sections. Moreover, it is also important to collect results related to the size of the duality gap obtained when using the weak formulation compared with the gap given by the stronger formulations proposed. This is particularly significant since in the literature, the computational results presented for the 2-level special cases of our general model refer only to the weak formulation. Another interesting issue is related with the strong formulations. Although the Lagrangian duals described in Secion 2.4.4 provide the same optimal bound, it is not clear which one is more efficiently optimized using the subgradient method. In addition, some results comparing the behavior of the subgradient method with the modified version are presented. Finally, it is important to collect results concerning the lower bounding methods described in Section 2.4.5, and the branch and bound algorithm proposed in the previous section. The way the profits of the model are generated seems to play an important role in the results obtained in any type of integer programming problems. Therefore, we considered two types of sample problems, denoted by *Euclidean* and *Random*. The main difference between these samples lies in the way the profits are generated. For the *Euclidean* sample they were generated according to the expression given in Section 2.4.1, namely $c_{ijk} = d_i\,(p_i - q_j - r_k - t\,D_{ijk})$. The following parameters were adopted

$$d_i \sim U(1,3); \quad p_i \sim U(100,200); \quad q_j \sim U(1,5); \quad r_k \sim U(1,5); \quad t = 0.1$$

where $U[a,b]$ denotes the integer uniform random distribution and the distances D_{ijk} are rounded Euclidean distances between randomly generated points in the grid $(0,100) \times (0,100)$. For the *Random* sample, the profits c_{ijk} were uniformly generated in $(100,400)$.

The operating fixed costs and the fixed costs of the facilities and depots were randomly generated in different intervals taking into account the number of clients. Thus, for 10 and 20 clients the fixed costs were generated as follows

$$f_j \sim U(700, 1000), \quad g_k \sim U(300, 800), \quad F_{jk} \sim U(100, 400),$$

while for 30 and 50 clients

$$f_j \sim U(2100, 3000), \quad g_k \sim U(900, 2400) \quad F_{jk} \sim U(300, 1200).$$

For each dimension we took samples of 5 uncorrelated test problems, so all the results presented correspond to averages.

In order to compare the behavior of our general model with the 2-level special cases we consider the following instances of these models. The test problems for the model proposed by Gao and Robinson in [59] were obtained by removing from the original test problems the fixed costs of the depots ($g_k = 0$). The instances of the 2-level uncapacitated facility location problem defined in [131] were obtained after removing all the operating fixed costs ($F_{jk} = 0$) from the original test problems.

All the programs were written in SUN Pascal and run on a Sun Sparc System 600 workstation. The code used to solve the min-cut problems is based on the highest-label preflow-push algorithm, see [2]. The SUN Pascal compiler was used with the default compilation options.

In order to optimize the Lagrangian duals ($\mathcal{DL}^{WF}$), ($\mathcal{DL}^{SF}$) and ($\mathcal{DL}^{BSF}$) we used the modified subgradient method described in Section 2.4.4. This choice was made after comparing the behavior of the modified subgradient method versus the usual method to optimize ($\mathcal{DL}^{SF}$) for the test problems. A collection of the main results of this test is contained in Table 2.1. Column *Problems* indicates the dimension of the test problems, where m corresponds to the number of clients, p to the number of facilities and q to the number of depots. Under the column *Euclidean* we only present the average number of iterations taken until the gap was closed, since for all the Euclidean examples there was no duality gap. Columns *ItSub* and *ItMod* contain respectively the average number of iterations of the classical subgradient method and of the modified method. The column *Random* contains the average duality gap for the test problems generated randomly. Similarly, columns *GSub* and *GMod* contain respectively the average gap obtained when using the classical subgradient method and the modified method.

Problems	Euclidean		Random	
(m, p, q)	ItSub	ItMod	GSub	GMod
$(10, 5, 5)$	28	12.0	0.87	0.87
$(20, 5, 5)$	27	5.2	3.01	2.77
$(20, 5, 10)$	35	9.6	2.82	2.69
$(30, 10, 10)$	34	8.0	3.24	2.70
$(30, 10, 15)$	35	9.0	9.24	8.22
$(50, 10, 10)$	37	14.0	8.75	8.59
$(50, 10, 15)$	38	15.0	10.89	10.42

Table 2.1: Variants of subgradient applied to $(\mathcal{DL}^{SF})$.

In the following tables, columns *dual* and *BLB* refer to the percent gaps obtained using the optimal value, $Z^\star$. For instance, for $(\mathcal{DL}^{WF})$ column *dual* refers to the duality gap, $\frac{\vartheta(\mathcal{DL}^{WF}) - Z^\star}{Z^\star} \times 100$, and column *BLB* refers to the gap between the best lower bound solution obtained during the optimization of $(\mathcal{DL}^{WF})$ and the optimal value. Column *It* refers to the number of iterations performed by the modified subgradient method and column *Sec* refers to the elapsed user time in seconds (given by the standard function `clock` of the SUN Pascal compiler) on the mentioned workstation. Finally, column *Greedy* refers to the average gap between the optimal value and the lower bound provided by the greedy heuristic.

The maximum number of iterations allowed in the modified subgradient method used to optimize the three duals, $(\mathcal{DL}^{WF})$, $(\mathcal{DL}^{SF})$ and $(\mathcal{DL}^{BSF})$ for the different models is as follows

	General	$g_k = 0$	$F_{jk} = 0$
iterations	550	500	500

At each node of the branch and bound the maximum number of iterations allowed to optimize $(\mathcal{DL}^{SF})$ depends on what occurred at the previous node. Hence, the maximum number of iterations at the beginning of the enumeration is 500, during a descent search 100 and after fathoming a node and returning to a pending one 200.

At each node, we use as the initial multipliers the last ones used at the previous node, following the strategy proposed by Fisher [50].

For all three models the *Euclidean* examples did not exhibit a duality gap for the strong formulations, as Tables 2.2, 2.3 and 2.4 show. On the other hand, the duality gap of the weak formulation appears to be very large. Interestingly, optimizing $(\mathcal{DL}^{BSF})$ appears to be more efficient.

Problems	$(\mathcal{DL}^{WF})$			$(\mathcal{DL}^{SF})$			$(\mathcal{DL}^{BSF})$		
(m, p, q)	dual	It	Sec	dual	It	Sec	dual	It	Sec
$(10, 5, 5)$	123.10	396.4	1.9	0.00	12.0	0.0	0.00	4.8	0.0
$(20, 5, 5)$	18.33	456.6	3.3	0.00	5.2	0.0	0.00	4.4	0.0
$(20, 5, 10)$	23.67	507.2	7.4	0.00	9.6	0.1	0.00	6.6	0.1
$(30, 10, 10)$	80.72	358.6	15.7	0.00	8.0	0.3	0.00	6.2	0.2
$(30, 10, 15)$	70.57	394.6	25.7	0.00	9.0	0.4	0.00	6.6	0.3
$(50, 10, 10)$	32.84	497.6	31.8	0.00	14.0	0.7	0.00	6.8	0.4
$(50, 10, 15)$	32.39	496.2	43.8	0.00	15.0	1.1	0.00	7.6	0.7

Table 2.2: Upper bounds for the *Euclidean* examples.

Problems	$(\mathcal{DL}^{WF})$			$(\mathcal{DL}^{SF})$			$(\mathcal{DL}^{BSF})$		
(m, p, q)	dual	It	Sec	dual	It	Sec	dual	It	Sec
$(10, 5, 5)$	39.03	380.8	0.9	0.00	6.8	0.0	0.00	5.6	0.0
$(20, 5, 5)$	10.69	399.6	1.7	0.00	8.0	0.0	0.00	5.0	0.0
$(20, 5, 10)$	14.30	396.2	3.0	0.00	7.0	0.1	0.00	6.2	0.1
$(30, 10, 10)$	40.94	360.4	7.9	0.00	11.6	0.3	0.00	16.0	0.4
$(30, 10, 15)$	36.97	372.8	11.6	0.00	6.8	0.2	0.00	6.2	0.2
$(50, 10, 10)$	18.85	376.4	13.9	0.00	7.0	0.3	0.00	6.6	0.3
$(50, 10, 15)$	19.21	397.0	20.9	0.00	9.2	0.5	0.00	9.8	0.6

Table 2.3: Upper bounds for the *Euclidean* examples $g_k = 0$.

For all three models the *Euclidean*, the greedy heuristic as well as the Lagrangian based heuristics always provided the optimal solution, and therefore we did not include the correspondent tables.

Problems	$(\mathcal{DL}^{WF})$			$(\mathcal{DL}^{SF})$			$(\mathcal{DL}^{BSF})$		
(m,p,q)	dual	It	Sec	dual	It	Sec	dual	It	Sec
$(10,5,5)$	90.56	332.2	1.5	0.00	10.6	0.0	0.00	4.6	0.0
$(20,5,5)$	20.42	397.2	4.9	0.00	9.4	0.1	0.00	5.8	0.0
$(20,5,10)$	22.92	408.8	9.5	0.00	12.0	0.2	0.00	7.4	0.1
$(30,10,10)$	72.91	492.2	21.6	0.00	12.8	0.5	0.00	7.6	0.3
$(30,10,15)$	67.92	455.0	29.6	0.00	21.2	1.1	0.00	11.2	0.5
$(50,10,10)$	31.03	322.6	21.2	0.00	15.6	1.0	0.00	8.4	0.5
$(50,10,15)$	31.73	317.8	30.4	0.00	18.2	1.6	0.00	9.2	0.7

Table 2.4: Upper bounds for the *Euclidean* examples $F_{jk} = 0$.

The results for the *Random* examples are quite different. In the case of the general
model, Table 2.5 indicates that the gains obtained by considering the stronger formulations are significant, achieving in some cases a reduction of the duality gap by
a factor of 10. The extra effort required to optimize $(\mathcal{DL}^{BSF})$ is not as substantial
as one would expect and is comparable to the effort needed to optimize $(\mathcal{DL}^{SF})$.

Problems	$(\mathcal{DL}^{WF})$			$(\mathcal{DL}^{SF})$			$(\mathcal{DL}^{BSF})$		
(m,p,q)	dual	It	Sec	dual	It	Sec	dual	It	Sec
$(10,5,5)$	54.29	403.6	1.9	0.87	152.8	0.5	0.87	148.2	0.6
$(20,5,5)$	21.96	308.8	2.6	2.77	550.0	3.8	2.76	550.0	4.7
$(20,5,10)$	22.81	379.0	5.9	2.69	550.0	6.4	2.73	550.0	8.0
$(30,10,10)$	60.61	363.6	16.0	2.70	357.0	10.8	2.80	353.6	13.1
$(30,10,15)$	80.88	380.0	29.0	8.22	459.4	20.0	8.54	455.2	24.9
$(50,10,10)$	40.43	376.0	25.2	8.59	550.0	29.0	8.59	550.0	36.1
$(50,10,15)$	40.53	362.8	46.0	10.42	550.0	40.6	10.79	550.0	50.7

Table 2.5: Upper bounds for the *Random* examples.

From Table 2.6 it appears that the quality of the lower bound solutions obtained
from the Lagrangian solutions when considering the strong formulations, (SF) and
(BSF) is quite good. Moreover, the performance of the greedy heuristic is, in this
case, less promising since the maximum observed gap was 7.87%.

Problems	$(\mathcal{DL}^{WF})$	$(\mathcal{DL}^{SF})$	$(\mathcal{DL}^{BSF})$	
(m, p, q)	BLB	BLB	BLB	Greedy
$(10, 5, 5)$	0.00	0.00	0.00	0.00
$(20, 5, 5)$	2.13	0.35	0.35	2.13
$(20, 5, 10)$	2.50	0.00	0.79	2.50
$(30, 10, 10)$	0.00	0.00	0.00	0.00
$(30, 10, 15)$	1.07	0.00	0.95	1.07
$(50, 10, 10)$	1.24	0.00	0.01	1.24
$(50, 10, 15)$	0.48	0.48	0.00	0.48

Table 2.6: Lower bounds for the *Random* examples.

Table 2.7 contains the results obtained by the branch and bound algorithms described in Section 2.4.6. Under column *On (y_j, z_k)* we have the results using the branching rule which considers first a (facility,depot) pair, creating four mutually exclusive descendant nodes: $(y_j = 0$ and $z_k = 0)$, $(y_j = 0$ and $z_k = 1)$, $(y_j = 1$ and $z_k = 0)$ and $(y_j = 1$ and $z_k = 1)$. On the other hand, column *On t_{jk}* contains the results obtained using a simple binary rule on t_{jk}. Finally, column *Nodes* refers to the average number of nodes evaluated by the branch and bound algorithm.

Problems	On (y_j, z_k)		On t_{jk}	
(m, p, q)	Nodes	Sec	Nodes	Sec
$(10, 5, 5)$	1.0	0.64	1.4	0.77
$(20, 5, 5)$	6.6	5.72	7.8	7.17
$(20, 5, 10)$	11.0	10.54	14.2	16.72
$(30, 10, 10)$	9.4	19.03	10.6	30.03
$(30, 10, 15)$	22.4	49.97	42.8	129.90
$(50, 10, 10)$	33.0	98.51	62.6	245.12
$(50, 10, 15)$	55.8	196.69	109.8	567.61

Table 2.7: *Random* examples, Branch and Bound.

The branch and bound algorithm seems to be able to solve quite efficiently the test problems where duality gaps exist. Moreover, it appears that the mixed branching rule performs much better than the simple binary rule on t_{jk}.

It is interesting to see how much "easier" the problem becomes after removing the fixed costs of the depots. One possible explanation for this phenomenon lies in the structure of the 2-echelon uncapacitated facility location problem. Observe that now, deciding if a given pair should operate together only depends on the fixed costs of the facility and the operating costs. Therefore, this problem seems to be "closer" to the uncapacitated facility location problem. Table 2.8 shows that for this type of problems, formulations (SF) and (BSF) are very "strong" in the sense that their linear relaxations reduce substantially the duality gap and in some cases they are enough to solve the problem. Hence, it seems important to consider the inclusion of the facets defined by (2.21) in the formulation of the 2-echelon uncapacitated facility location problem. We see again that the effort made to optimize $(\mathcal{DL}^{BSF})$ is very similar to the one used to optimize $(\mathcal{DL}^{SF})$.

Problems	$(\mathcal{DL}^{WF})$			$(\mathcal{DL}^{SF})$			$(\mathcal{DL}^{BSF})$		
(m, p, q)	dual	It	Sec	dual	It	Sec	dual	It	Sec
$(10, 5, 5)$	20.18	442.0	1.0	0.00	151.6	0.4	0.00	86.6	0.2
$(20, 5, 5)$	4.77	468.2	1.9	0.00	52.4	0.3	0.00	33.6	0.2
$(20, 5, 10)$	6.19	456.2	3.3	0.00	106.8	0.8	0.00	68.8	0.6
$(30, 10, 10)$	21.09	460.4	8.9	0.21	166.0	3.4	0.21	146.6	3.2
$(30, 10, 15)$	22.30	456.4	13.4	0.20	324.8	9.1	0.22	313.8	9.4
$(50, 10, 10)$	10.86	462.6	15.7	0.36	322.2	11.3	0.35	315.4	12.1
$(50, 10, 15)$	10.87	474.4	23.0	0.29	367.8	18.9	0.25	362.2	17.0

Table 2.8: Upper bounds for the *Random* examples $g_k = 0$.

As Table 2.9 shows the performance of the greedy heuristic is again not too promising (the maximum observed gap was 15.43%). However, the quality of the lower bound solutions obtained from any of the Lagrangian solutions is higher than that of the greedy heuristic. Again, better solutions are derived from the "stronger" formulations, (SF) and (BSF).

Table 2.10 contains the results obtained when using the branch and bound algorithms described in Section 2.4.6. Under column On (y_j, t_{jk}) we have the results using the branching rule that branches first on the variables y_j until there are no more free facilities and the current node cannot be fathomed.

Problems	$(\mathcal{DL}^{WF})$	$(\mathcal{DL}^{SF})$	$(\mathcal{DL}^{BSF})$	
(m, p, q)	BLB	BLB	BLB	Greedy
$(10, 5, 5)$	2.26	0.04	0.00	4.08
$(20, 5, 5)$	0.38	0.00	0.00	3.69
$(20, 5, 10)$	2.62	0.00	0.00	3.53
$(30, 10, 10)$	2.37	0.00	0.00	2.37
$(30, 10, 15)$	4.85	0.00	0.00	4.85
$(50, 10, 10)$	2.33	0.35	0.16	2.94
$(50, 10, 15)$	0.50	0.00	0.00	0.50

Table 2.9: Lower bounds for the *Random* examples $g_k = 0$.

Problems	On (y_j, t_{jk})		On t_{jk}	
(m, p, q)	Nodes	Sec	Nodes	Sec
$(10, 5, 5)$	0.2	0.49	0.2	0.58
$(20, 5, 5)$	0.0	0.34	0.0	0.38
$(20, 5, 10)$	0.0	1.07	0.0	1.13
$(30, 10, 10)$	2.8	6.54	0.6	5.45
$(30, 10, 15)$	8.6	19.40	4.2	17.21
$(50, 10, 10)$	13.2	38.57	4.6	32.70
$(50, 10, 15)$	10.4	47.65	3.4	39.90

Table 2.10: *Random* examples $g_k = 0$, Branch and Bound.

Only then branching on (the remaining) t_{jk} is initiated. On the other hand, column *On t_{jk}* contains the results obtained using a simple binary rule on t_{jk}. The performance of the branch and bound algorithm is promising, and so we can expect to solve larger instances than those presented by Gao and Robinson [59, 60]. Moreover, it appears that in this case the binary rule on t_{jk} performs better than the mixed branching rule. Observe also that Gao and Robinson [59] obtain an average duality gap of 0.16% by approximating the linear bound given by the weak formulation (WF) for their test problems. Clearly, no direct comparison between these results and ours can be made since our test problems present much larger duality gaps for the weak formulation. Moreover, since we optimize the Lagrangian dual instead of using heuristics to approximate its bound, our upper bounds would be smaller than those obtained with Gao and Robinson's strategy.

The instances of the 2-level uncapacitated facility location problem, i.e. those with $F_{jk} = 0$, appear to be surprisingly "harder" than the original problems. This phenomenon may be explained by the link between the two types of entities established by the presence of positive F_{jk}. In fact, the existence of operating (infrastructure) fixed costs can be seen as an extra "help" to differentiate among possible operating pairs.

Problems	$(\mathcal{DL}^{WF})$			$(\mathcal{DL}^{SF})$			$(\mathcal{DL}^{BSF})$		
(m, p, q)	dual	It	Sec	dual	It	Sec	dual	It	Sec
$(10, 5, 5)$	63.53	321.0	1.5	19.43	500.0	1.8	18.68	500.0	2.5
$(20, 5, 5)$	25.79	359.4	2.7	12.17	500.0	3.7	11.71	500.0	4.6
$(20, 5, 10)$	28.44	333.2	4.5	13.12	500.0	6.4	12.67	500.0	8.3
$(30, 10, 10)$	70.92	310.6	12.5	26.56	500.0	17.3	25.87	500.0	23.2
$(30, 10, 15)$	84.86	292.6	17.8	35.02	500.0	24.6	34.16	500.0	32.3
$(50, 10, 10)$	46.98	296.6	18.3	25.18	500.0	29.5	24.43	500.0	37.1
$(50, 10, 15)$	44.59	305.4	28.4	24.02	500.0	41.6	23.35	500.0	52.3

Table 2.11: Upper bounds for the Random examples $F_{jk} = 0$.

Table 2.11 shows that the strong formulations could only reduce the duality gap of the weak formulation by half. Also for this type of problems solving $(\mathcal{DL}^{BSF})$ seems to be worthwhile.

Problems	$(\mathcal{DL}^{WF})$	$(\mathcal{DL}^{SF})$	$(\mathcal{DL}^{BSF})$	
(m, p, q)	BLB	BLB	BLB	Greedy
$(10, 5, 5)$	0.00	0.00	0.00	0.00
$(20, 5, 5)$	0.91	0.01	0.22	0.91
$(20, 5, 10)$	0.00	0.00	0.00	0.00
$(30, 10, 10)$	0.80	0.80	0.05	0.80
$(30, 10, 15)$	3.60	2.25	3.60	3.60
$(50, 10, 10)$	0.68	0.19	0.40	0.68
$(50, 10, 15)$	0.55	0.55	0.55	0.55

Table 2.12: Lower bounds for the Random examples $F_{jk} = 0$.

Table 2.12 indicates that the behavior of the greedy heuristic is not very good (the maximum observed gap was 8.0%). It also appears that the Lagrangian solutions obtained from the "stronger" formulations are better than the ones obtained by the greedy heuristic.

Problems	On (y_j, z_k)	
(m, p, q)	*Nodes*	*Sec*
$(10, 5, 5)$	13.0	4.69
$(20, 5, 5)$	31.0	15.26
$(20, 5, 10)$	56.2	38.90
$(30, 10, 10)$	67.2	103.91
$(30, 10, 15)$	149.8	292.58
$(50, 10, 10)$	203.4	473.20
$(50, 10, 15)$	404.2	1245.66

Table 2.13: *Random* examples $F_{jk} = 0$, Branch and Bound.

Table 2.13 contains the results obtained when using the branch and bound algorithm, described in Section 2.4.6. Under column *On* (y_j, z_k) we have the results obtained using the branching rule which branches on a (facility,depot) pair (y_j, z_k). As mentioned in Section 2.4.6, the maximum number of nodes potentially created by our branching rule is smaller, for $p, q \geq 2$ than the corresponding number from the binary branching rule on y_j and z_k, presented by Tcha and Lee [131]. Moreover, branching on t_{jk} (artificially created) corresponds to our branching rule. The poor quality of the lower and upper bounds explains the bad performance of the branch and bound algorithm. No meaningful comparisons can be made with the results presented by Tcha and Lee [131] because their branch and bound algorithm includes an incorrect node simplification method, see [18], and so their results are not worthy: they can prematurely discard important nodes and thus prune branches containing optimal solutions.

2.5 Conclusions

Generalizations of the classical uncapacitated facility location problem considering more levels of facilities to be located, became more important in the last two decades.

In particular, the extension of the basic model to the 2-level case seems to be suitable for many real life applications. Within the uncapacitated 2-level case, our model explores the relations among the facilities to be located by considering simultaneously three types of fixed costs. As a result, the model generalizes two known models in the literature namely the 2-echelon uncapacitated facility location problem and the 2-level uncapacitated facility location problem. Notice also, that since the 2-echelon uncapacitated facility location problem generalizes the multi-activity uncapacitated facility location problem our model also covers this case.

The computational results indicate that our problem is "nastier" than the classical uncapacitated facility location problem. This also appears to be the case for the 2-echelon uncapacitated facility location problem and the 2-level uncapacitated facility location problem. Moreover, due to the way the test problems where generated (instances of the special models correspond to the basic test problems conveniently adapted) it seems that for similar data each of these 2-level problems has distinct behavior.

Although Proposition 2.4.1 establishes that the linear relaxation of formulation (WF) is never better than (BSF) and (SF), the computational results show that $(\overline{WF})$ is surprisingly "weak". Therefore, extensions of Erlenkotter's dual adjustment method approximating the optimal value of the $(\overline{WF})$ are not as promising as they are for the uncapacitated facility location problem. This may be explained by the fact that Erlenkotter's method uses the strong formulation of the uncapacitated facility location problem which includes facets (2.2). Actually, the inclusion of facets (2.21) and (2.22) in both formulations (SF) and (BSF) produces significantly smaller gaps. These results suggest the importance of deriving more valid inequalities for this problem. This appears to be particularly relevant for the 2-level uncapacitated facility location problem. The "stronger" formulation proposed in Section 2.4.2 compared to the original formulation of Tcha and Lee [131] reduces the duality gap significantly. Recently, Aardal *et al.* [1] have proposed new valid inequalities in order to strengthen the "stronger" formulation for the 2-level uncapacitated facility location problem. However, the practical effect of these new inequalities is not known yet.

The modified subgradient proposed in Section 2.4.4 to optimize Lagrangian duals satisfying the integrality property appears to be more efficient than the classical subgradient method. Moreover, this modified method is still as simple as the original one.

Another interesting issue is to compare the behavior of the modified subgradient method when solving $(\mathcal{DL}^{BSF})$ and $(\mathcal{DL}^{SF})$. Although the Lagrangian relaxation presented for the (BSF) requires more effort to be solved than the Lagrangian relaxation for (SF), this effort is usually compensated by the total number of iterations.

Contrary to the special case of the uncapacitated facility location problem, the greedy heuristic appears to perform poorly. A possible explanation for this is the fact that the problem is no longer submodular. This property, if valid, would indeed imply a worst-case bound for the performance of the greedy heuristic, see [51].

Finally, different ways of generating the profits of assigning the clients appear to be important in the behavior of the problem. It is interesting to note that when the profits are generated according to the formula proposed in Section 2.4.1 (reasonable for most real life applications), the linear relaxation of the stronger formulation gives nearly always the optimal solution. For the uncapacitated facility location problem the same type of behavior was also observed and is supported by probabilistic results, see [39]. The way the test problems are generated is also extremely important when relating computational results obtained using different approaches. Moreover, comparing results produced by different test problems can lead to erroneous conclusions. For instance, from the computational experience reported by Gao and Robinson [59, 60] it appears that the (WF) formulation provides very good bounds for the 2-echelon uncapacitated facility location problem. On the other hand, our two sets of test problems point in the opposite direction. The main difference lies in the fact that the test problems used by Gao and Robinson present average duality gaps for the (WF) of about 0.16% while for our Euclidean generated problems the average gaps for the same formulation are about 25% and for the random examples about 17%. Also in the case of the 2-level uncapacitated facility location problem, no meaningful comparison can be made with the results of Tcha and Lee [131] since not only they do not report the duality gaps of their test problems, but also the simplification rule used in their work is not correct.

Three

Location Models and
Fractional Programming

The discrete location models discussed in the previous chapter distinguish the location decisions by the total net profit. The total net profit measures the gains of a certain location decision, i.e. the difference between the sum of the profits associated with serving each client via certain facilities and the fixed costs associated with those facilities. This traditional criterion assumes that the profits and fixed costs are available and correctly computed. In operations research the subject of finding these parameters is usually left to economists and finance consultants. However, as we shall see, it may be decisive to understand how economists obtain the data in order to correctly apply the models studied in the previous chapter. Therefore, we will briefly introduce some of the most popular economical concepts, which are discussed in greater detail by Brealey and Myers [30].

Observe first that a location decision can be interpreted as a project which takes place during a certain time period. In fact, while the fixed costs are "paid" at the beginning of the project, the so-called profits associated with allocating clients to facilities correspond to the sum of all the cash-flows during the time period in question. Assuming that these cash-flows can be estimated it is left to determine how they should be considered as profits in a meaningful way. This last issue is justified by the maxim in economics and finance: *a dollar today is worth more than a dollar tomorrow*. Moreover, it is also important to evaluate what would have been the cash-flows if the initial investment were spent on another type of project with similar features. In order to quantify this last effect, usually designated by the *discount factor*, it is required to know the *opportunity cost of capital* or expected *rate of return*, which reflects the return that one would obtain by investing in similar projects. Assuming that the discount factor per time period is available then the *present value* of a project is given by the sum of the cash-flows weighted by the corresponding discount factor during the duration of the project. Clearly, the profitability of a project can be measured by subtracting the initial investment from the present value. This measure is usually known as the *net present value*, and considers a project to be worthwhile to accept whenever its net present value is positive. Therefore, the traditional criterion in location science of maximizing the total net profit is economically justified when computing the profits according to their present value.

Another approach to determine the profitability of a project is given by the *profitability index*. The profitability index is defined by the ratio between the present value of a project and the required investment made. Clearly, a project is worthwhile whenever its profitability index is greater than one, which corresponds to having its net present value greater than zero. In order to clarify these concepts we will consider the following small example. Let A and B be two projects taking place during the same time period (three unit periods) and suppose that the expected cash-flows and the necessary investments are known. Assuming that the present values of these projects are computed using the discounted cash-flow formula, see [30], Figures 3.1 and 3.2 show the values of the net present values (npv), respectively, the profitability indexes (pi) of projects A and B according to different opportunity costs of capital (occ).

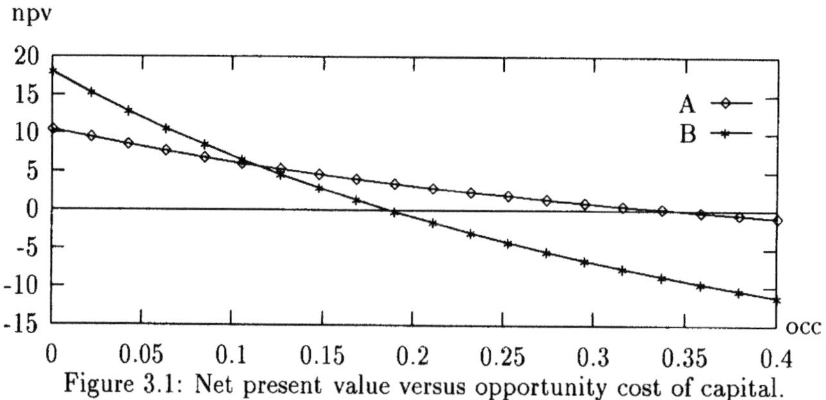

Figure 3.1: Net present value versus opportunity cost of capital.

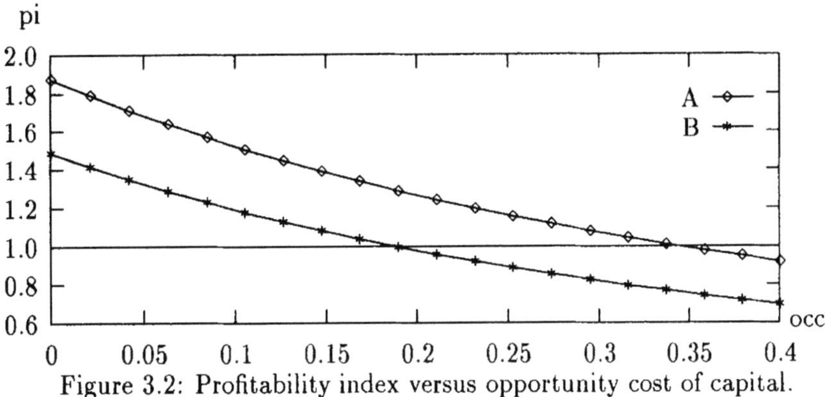

Figure 3.2: Profitability index versus opportunity cost of capital.

This small example clearly shows that finding the project which maximizes each of these rules does not always provide the same answer. Hence, it is important to characterize the different situations for which either the profitability index rule or the net present value rule are more suitable. Maximizing the net present value is the most popular rule used by financial managers. However, there are situations where it is important to consider the maximization of the profitability index. Selecting the project which maximizes the profitability index, instead of the one with the highest net present value, can be a "safe" decision. In fact, if the opportunity cost of capital is hard to estimate and subject to high variations (for instance, due to an unstable economical situation) this affects more the net present value than the profitability index. Figures 3.1 and 3.2 show that if the opportunity cost of capital is assumed to be 7.5% then using the net present value rule project B would be selected. However, if the opportunity cost of capital is in fact 20% project B is no longer worthwhile. Therefore, selecting the project which maximizes its profitability index yields a less risky decision in case the opportunity cost of capital is hard to estimate.

Also, under a limited investment budget it becomes more important to select a project which will maximize the profitability index rather than the net present value, assuming the existence of other projects with positive net present values as the next situation illustrates. Suppose that a firm has a limited budget of 120 units. This firm has the possibility of investing in a location project L. Maximizing the net present value over all the possible location alternatives yields a solution L_1 with a net present value of 214 for which an investment of 110 is required. The associated profitability index is 3.1. On the other hand, maximizing the profitability index originates a solution L_2 requiring an investment of 50 with a profitability index of 4.2 and associated net present value of 164. Assume now that there is an extra investment possibility given by the project P requiring an initial investment of 55 units, yielding a net present value of 123 units and a profitability index of 3.2. In this case, by investing in L_2 (the location solution which maximizes the profitability index) and in project P, the firm will obtain a larger total net present value than by considering only the location project with the highest net present value, i.e. L_1[1].

The above discussion motivated the analysis of discrete location models considering the maximization of the profitability index (or rate of return). Due to this, some references to location models in which the profitability index is maximized have appeared recently in the literature (Barros *et al.* [13], Myung and Tcha [95], Revelle and Laporte [113] and Hansen *et al.* [68]). Clearly, such problems belong now to the field of integer fractional programming. Due to the nonlinear nature of the objective function, this class of integer programming problems appears to be more difficult to solve than its counterpart with a linear type objective function. However, as we shall see in Section 3.2, there exist some special classes of location problems with ratios as the objective function that can be trivially solved. These results are easily derived using basic results of fractional programming and therefore, we will start this section by presenting a brief survey on fractional programming and its techniques.

3.1 Fractional Programming

Fractional programming describes a particular class of nonlinear programming dealing with constrained ratio optimization problems. The term *fractional programming* has been adopted in the literature simplifying *fractional functionals programming* suggested by Charnes and Cooper [34]. Occasionally, the term *hyperbolic programming*, introduced by Martos [90], is also used.

A typical application occurs in management analysis, where the efficiency of a system must be maximized. This can be expressed by considering minimizing the ratio between the amount of resource wasted and used on the production plan. Also in

---

[1] We assume that the cash-flows are nonnegative and that the location project and project P are not dependent on each other.

the field of economical analysis, more specifically for portfolio selection problems, it is required to maximize the ratio between the expected return and risk of the investments. Applications of fractional programming can also be found, for instance in information theory, stochastic programming, numerical analysis, game theory and maintenance. A more detailed description of these applications can be found in Craven [40].

Besides the mentioned examples, the field of location analysis also yields some interesting applications of fractional programming as we shall see in Section 3.2. Also, a rather surprising application of fractional programming can be found in continuous location. In fact, Elzinga *et al.* [47] show that the dual problem of a *minimax multi-facility location problem* with Euclidean distances corresponds to a concave fractional programming problem. Observe that this relation is not obvious at first sight due to the format of this location problem which corresponds to locating new facilities among existing ones with the objective of minimizing the maximum weighted Euclidean distance among all facilities.

We will begin this section by reviewing the usual fractional programming problems and only then we will give a brief survey on the important and usually more neglected field of integer fractional programming. For a survey and an up-to-date (until late 1993) exhaustive bibliography on fractional programming see [124].

3.1.1 Continuous Fractional Programming

Continuous fractional programming problems are usually defined in the following way. Let $\mathcal{X} \subset I\!R^m$ and $f, g : \mathcal{C} \longrightarrow I\!R$ be continuous functions with $\mathcal{C}$ an open set containing $\mathcal{X}$. If additionally $g(x) > 0$ for every $x \in \mathcal{X}$, a fractional program is given by

$$\sup_{x \in \mathcal{X}} \frac{f(x)}{g(x)}. \qquad (P)$$

In the remaining it is always assumed that an optimal solution to (P) exists. A sufficient condition to achieve this is given by assuming that the feasible set $\mathcal{X}$ is compact.

An important class of fractional problems is given by *concave fractional programming* problems. For these fractional programming problems the feasible set $\mathcal{X}$ is convex and given by $\mathcal{X} := \{x \in \mathcal{S} : h(x) \leq 0\}$ with $h : \mathbb{R}^m \longrightarrow \mathbb{R}^r$ a convex vector-valued function and $\mathcal{S} \subseteq \mathcal{C}$ a convex set. Additionally, the functions $f, g : \mathcal{C} \longrightarrow \mathbb{R}$ are respectively concave and convex on $\mathcal{S}$ and g is positive on $\mathcal{X}$. Moreover, either the function f is nonnegative on $\mathcal{X}$ or g is affine.

The class of concave fractional programs is particularly important since the ratio is a semistricly quasiconcave function on $\mathcal{X}$ and this property ensures that a local maximum of (P) is a global maximum, see [6]. Furthermore, if either the numerator is strictly concave or the denominator is strictly convex on $\mathcal{X}$, then the maximum is unique.

The most well-known and studied class of concave fractional programming problems is probably the *linear fractional programming* class. As the name suggests, these fractional programs assume that both functions f and g are affine and that $\mathcal{X}$ is a convex polyhedron. For this special case, it follows that the objective function is also quasiconvex and this implies that any optimal solution will be attained at an extreme point of the polyhedron.

Due to the mentioned characteristics, concave fractional programming and its special case, linear fractional programming, have been extensively studied in the literature giving rise to four basic classes of solution procedures to tackle these problems. The most immediate approach corresponds to directly solving these problems. In case of differentiability, the objective function of a concave fractional program is pseudoconcave and hence a solution of the Karush-Kuhn-Tucker optimality conditions is a global maximum, see [6]. Therefore, several methods used to optimize concave problems also solve this particular class of concave fractional programs with pseudoconcave objective functions, like for instance the gradient-type method of Frank and Wolfe, see Martos [91].

Another interesting property of concave fractional programs arises from the upper subdifferentiability, of the objective function. As a result, the cutting plane algorithm of Plastria [106] can be directly applied to concave fractional programs. This idea was successfully explored by Boncompte and Martínez-Legaz [28].

In case of linear fractional programming with a bounded feasible set these problems can be solved using an adaptation of the simplex method. This simplex-like procedure, described in Martos [90, 91], generates a sequence of adjacent extreme points of the feasible set $\mathcal{X}$ with increasing objective values. Also, Anstreicher [4] shows that Karmarkar's projective algorithm is fundamentally an algorithm for linear fractional programming on the simplex. However, there exist some cases where solution procedures available for concave programming cannot be used directly to solve concave fractional programs as shown in [91]. Hence, it seems reasonable to transform the concave fractional program into an equivalent concave program.

Theorem 3.1.1

A concave fractional program (P) can be reduced to the concave program

$$\sup \left\{ t\, f(t^{-1} y) : t\, h(t^{-1} y) \leq 0,\, t\, g(t^{-1} y) \leq 1,\, t^{-1} y \in \mathcal{S},\, t > 0 \right\}. \qquad (P')$$

applying the transformation

$$y = x\, t \ \text{ and } \ t = \frac{1}{g(x)}. \qquad (3.1)$$

Moreover, if $(y^\star, t^\star)$ solves (P') then $x^\star = {t^\star}^{-1} y^\star$ solves (P).

The proof of this result due to Schaible can be found in [121]. Transformation (3.1) was initially suggested by Charnes and Cooper [34] for linear fractional programming. In this special case, (P') corresponds to a linear programming problem. Moreover, Wagner and Yaun [132] showed that solving the transformed problem (P') via the simplex method is algorithmically equivalent to applying the simplex-like algorithm of Martos [90] to the original problem. Clearly, the use of this transformation depends mostly on the structure of the involved functions. Examples of the application of this approach can be found in [123].

The above two approaches aim to solve the primal concave fractional problem or a transformation of this problem. An alternative approach consists of constructing the associated dual program and solving it. However, the basic duality concepts for concave programming cannot be directly applied to concave fractional programming. This suggests considering the concave problem (P') instead of (P). In this case, using classical results from Lagrangian duality the following dual fractional problem is derived

$$\inf_{z \geq 0} \left\{ \sup_{x \in \mathcal{S}} \frac{f(x) - z^{\mathsf{T}} h(x)}{g(x)} \right\}. \qquad (D)$$

In order to construct this dual problem of (P) we need to impose the stronger condition that $g(x) > 0$ for all $x \in \mathcal{S}$. Schaible [120, 121] discusses duality results relating this dual problem with (P') and the original problem. In fact, there exists an abundant collection of references on duality for concave fractional programming and in particular for linear fractional programming. Some of these references [21, 25, 78, 120, 121] discuss alternative duals for (P) or derive (D) using other approaches. Nevertheless, the use of duality in fractional programming has never been oriented towards the construction of efficient computational tools, except for linear fractional programming. In this special case, Bitran and Magnanti [25] use duality results to perform sensitivity analysis.

One of the most popular strategies for fractional programming is the parametric approach which considers the class of optimization problems associated with (P) given by

$$\sup_{x \in \mathcal{X}} \{f(x) - \lambda g(x)\} \qquad (P_\lambda)$$

with $\lambda \in \mathbb{R}$. Observe that for a concave fractional program the associated so-called parametric problem is a concave problem, while for linear fractional programming it corresponds to a linear programming problem. Assuming additionally that $\mathcal{X}$ is a compact set it follows since the function $x \longmapsto \frac{f(x)}{g(x)}$ is finite-valued and continuous on $\mathcal{X}$ that (P) is solvable, i.e. there exists some $x \in \mathcal{X}$ where the maximum of (P) is achieved. By a similar argument it also follows that (P_λ) is solvable.

To introduce the next lemma, relating (P) and (P_λ) we will consider the function $F : \mathbb{R} \longrightarrow \mathbb{R}$ given by

$$F(\lambda) := \max_{x \in \mathcal{X}} \{f(x) - \lambda g(x)\} .$$

Lemma 3.1.1 ([45])

If $\mathcal{X}$ is a compact set then

(a) *The function $F : \mathbb{R} \longrightarrow \mathbb{R}$ is convex, continuous and strictly decreasing.*

(b) *The optimal value $\lambda_\star$ of (P) is finite and $F(\lambda_\star) = 0$.*

(c) *$F(\lambda) = 0$ implies $\lambda = \lambda_\star$.*

(d) *The optimal solution set of $(P_{\lambda_\star})$ equals the optimal solution set of (P).*

From the above lemma it follows that solving (P) is equivalent to finding the unique root of the nonlinear univariate equation $F(\lambda) = 0$. Hence, it is essential to analyze the parametric function F. The following lemma describes some of the most important properties of this function.

Lemma 3.1.2

If $\mathcal{X}$ is a compact set the subgradient set of F at λ is given by

$$\partial F(\lambda) = \left[\min_{x \in \mathcal{X}(\lambda)} \{-g(x)\} , \max_{x \in \mathcal{X}(\lambda)} \{-g(x)\} \right],$$

with $\mathcal{X}(\lambda)$ denoting the set of optimal solutions of the parametric problem (P_λ), i.e., $\mathcal{X}(\lambda) := \{x \in \mathcal{X} : F(\lambda) = f(x) - \lambda g(x)\}$. Moreover, if $\lambda' \leq \lambda''$ then $g(x') \geq g(x'')$ for all $x' \in \mathcal{X}(\lambda')$ and $x'' \in \mathcal{X}(\lambda'')$.

Proof: The characterization of the subgradient set of F is a special case of Theorem 7.2 of [118] and therefore it is omitted. Moreover, an easy proof of this result is given in [14]. The proof of the remaining result can be found in [122].   □

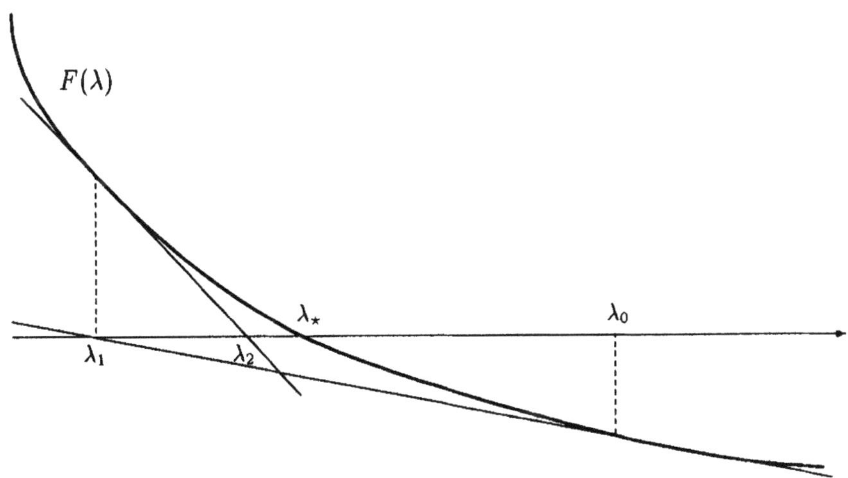

Figure 3.3: Geometric interpretation of Dinkelbach's algorithm.

From Lemma 3.1.2 it is clear for $x_{\lambda'}$ an optimal solution of $(P_{\lambda'})$ that the line given by $y = f(x_{\lambda'}) - \lambda g(x_{\lambda'})$ is tangent to $F(\lambda)$ at $\lambda = \lambda'$. Moreover, this line is always below the graph of the function F, see Figure 3.3. Thus, computing the root of $F(\lambda) = 0$ can be done by usual numerical methods, like Newton's algorithm or the bisection method.

Isbell and Marlow [76] applied Newton's algorithm to linear fractional programs and this approach was later generalized by Dinkelbach [45] for nonlinear fractional programs. Actually, this application of Newton's algorithm to fractional programming is usually known as Dinkelbach's algorithm. A description of this method is given in Algorithm 3.1.

$Step$ 0. Take $\boldsymbol{x}_0 \in \mathcal{X}$, compute $\lambda_1 := \dfrac{f(\boldsymbol{x}_0)}{g(\boldsymbol{x}_0)}$ and let $k := 1$;

$Step$ 1. Determine $\boldsymbol{x}_k := \arg \max_{\boldsymbol{x} \in \mathcal{X}} \left\{ f(\boldsymbol{x}) - \lambda_k g(\boldsymbol{x}) \right\};$

$Step$ 2. **If** $F(\lambda_k) = 0$

> **Then** $\boldsymbol{x}_k$ is an optimal solution of (P) with value λ_k and **Stop**.
> **Else GoTo** $Step$ 3;

$Step$ 3. Let $\lambda_{k+1} := \dfrac{f(\boldsymbol{x}_k)}{g(\boldsymbol{x}_k)}$, let $k := k + 1$, and **GoTo** $Step$ 1.

Algorithm 3.1: Dinkelbach's algorithm.

It is easy to show that the sequence λ_k, $k \geq 1$ generated by Dinkelbach's algorithm is increasing and satisfies $\lim_{k \uparrow \infty} \lambda_k = \lambda_\star$ with $\lambda_\star = \vartheta(P)$, see [45, 122]. Using the convexity of the parametric function F and the fact that the mapping $\lambda_k \longmapsto \partial F(\lambda_k)$ is upper semicontinuous it is easy to prove that the Dinkelbach algorithm has a superlinear convergence rate, see [41]. Schaible [122] discusses these results and also shows that if the function g is continuously differentiable on $\mathcal{X}$ then the algorithm has a quadratic convergence rate.

As we have seen Dinkelbach's algorithm constructs a sequence of points approximating the optimal value $\vartheta(P)$ from below. However, it is also possible to simultaneously provide an upper bound on the optimal value. Due to the convexity of the parametric function F we have for all λ that

$$F(\lambda) \geq F(\lambda_\star) + (\lambda - \lambda_\star)g^\star$$

with $g^\star \in \partial F(\lambda_\star)$ and $\lambda_\star$ the optimal value of (P). From Lemma 3.1.2 it follows that $-g^\star \geq \delta := \min_{x \in \mathcal{X}} g(x) > 0$, and hence by Lemma 3.1.1 we have

$$\lambda \leq \lambda_\star \leq \lambda + \frac{F(\lambda)}{\delta}. \tag{3.2}$$

This relation justifies stopping the algorithm whenever $F(\lambda_k) \leq \varepsilon \delta$, with ε the required precision. Observe that this stopping rule ensures that $\lambda_k \leq \lambda_\star \leq \lambda_k + \varepsilon$. Moreover, due to the fact that the sequence $\{\lambda_k\}_{k \geq 1}$ is increasing we also obtain that

$$\lambda_\star - \varepsilon \leq \lambda_{k+1} = \frac{f(x_k)}{g(x_k)} \leq \lambda_\star.$$

Dinkelbach's algorithm is extremely popular not only due to its simplicity but also because it can be applied to all types of fractional programs. Clearly, the computational efficiency of Dinkelbach's method depends mostly on the evaluation of the function F at each iteration point. Thus, this algorithm is particularly well suited for concave and linear fractional programming. In these particular cases, it is important to relate this approach with solving the transformed problem (P'). In this case, a single concave problem must be solved, while Dinkelbach's algorithm requires the solution of several concave problems. Hence, it appears to be more efficient to solve the transformed problem. However, for certain types of problems, it is possible to take advantage of the structure of the functions involved in the ratio when using the parametric approach. For instance, if f and g are quadratic functions the resulting parametric problem is also a quadratic problem while the transformed problem has a nonquadratic objective function and a (possibly nonlinear) extra constraint. Moreover, the number of parametric problems that has to be solved is relatively small in practice, due to the superlinear convergence rate of this algorithm.

Although this parametric approach is largely accepted, from a theoretical point of view it appears to be an ad-hoc procedure. This is mainly due to the surprising relation between (P) and (P_λ) established in Lemma 3.1.1. Sniedovich [129] showed that Dinkelbach's algorithm corresponds to a classical method of mathematical programming. In fact, fractional programming problems can be viewed as pseudolinear problems and therefore solvable using first-order necessary and sufficient optimality conditions, see [129].

The literature is quiet abundant in variants of the Dinkelbach algorithm. In particular, Schaible [122] pointed out the importance of a "good" starting point for

Dinkelbach's algorithm. Accordingly, it is proposed to first compute a sequence of improving upper and lower bounds using an efficient section method. When a set of lower and upper bounds is found within a prescribed tolerance, Dinkelbach's algorithm starts using the just found lower bound. Notice than in case of concave and linear fractional programming the correspondent dual problem (D) can be used to derive upper bounds. In fact, the value of any feasible solution of the dual problem is an upper bound on the optimal value.

Recently, Pardalos and Phillips [103] proposed another variant of Dinkelbach's algorithm that generates simultaneously a sequence of lower and upper bounds on the optimal value. The convergence rate of this algorithm remains superlinear and is considered one of the most efficient at the present. Their algorithm is somewhat similar to the Dinkelbach's variant introduced by Schaible [122]. The basic difference lies in the fact that Pardalos and Philips algorithm updates continuously both the lower and upper bounds. Moreover, their approach can be used to solve some general fractional programs like the ratio of quadratic functions or the ratio of functions such that the associated parametric function is partially separable. Notice, that in the general case (P) may have several local maxima that are different from the global maximum, and thus finding its optimal solution is a "difficult" problem.

Another parametric like approach consists in constructing a succession of intervals containing the optimum value with decreasing amplitudes. Ibaraki [75] presents several of these interval type algorithms, which combine Dinkelbach's method with various search techniques. According to Ibaraki these algorithms, seem to perform slightly better than Dinkelbach's original algorithm. These results and an extensive discussion of the most efficient interval type algorithms can be found in [75].

Besides the mentioned classical approaches to tackle (P), lately new techniques have emerged from the interior point field for linear and concave fractional programming problems. The first reference on how to apply interior point techniques to linear fractional programming is due to Anstreicher [4]. Freund and Jarre proposed in [54] an interior point algorithm to solve directly the class of fractional programming programs (P), with a convex feasible set and a fraction of linear functions. This algorithm converges at a polynomial rate that depends on the self-concordance parameter of a barrier function associated with the feasible set. Another strategy consists in applying a standard interior point algorithm to the transformed problem (P'), which is a concave problem. In this case, Freund *et al.* [56] show that it is possible to construct the "best possible" self-concordance barrier for the feasible set of the transformed problem (P'). This interesting result enables also to prove that the rate of convergence is essential the same as the one in [54]. More recently, Nemirovsky [98, 99] proposed another interior point algorithm for fractional linear programming, the so-called method of analytic centers, that has a polynomial time complexity bound.

So far we have assumed that the feasible set of the fractional programming problem was compact. In practice this assumption may be too restrictive as shown in [10]. However, fractional programming problems with noncompact feasible sets yield, besides the existence or not of an optimal feasible solution, problems when applying usual fractional programming solution procedures, as discussed by Mond [94] and Cambini *et al.* [32]. In particular, Wagner and Yuan [132] show that, while for linear fractional programming with a compact feasible set the methods of Isbell and Marlow [76], Martos [90] and Charnes and Cooper [34] are equivalent for the noncompact case this equivalence fails. Moreover, as shown by Mond [94], the methods of Isbell and Marlow [76] and Martos [90] may fail to recover an optimal feasible solution. Recently Cambini and Martein [31] and Cambini *et al.* [32] have proposed modifications of respectively the methods of Isbell and Marlow and Charnes and Cooper for linear fractional programs with a noncompact feasible set. Finally, even when an optimal solution of a fractional programming problem exists applying Dinkelbach's algorithm to find it may pose difficulties. Since this algorithm uses iteratively an optimal feasible solution of the parametric problem for appropriate values of $\lambda < \lambda_\star$ one presumes that these solutions indeed exist. This holds for compact feasible sets but it is not clear in advance whether it also holds for noncompact feasible sets. Therefore, in order to apply Dinkelbach's algorithm one has to check in the first place if the problem has an optimal solution. Secondly, an interval containing the optimal solution to the problem has to be found on which the parametric problem has always an optimal feasible solution. An example of application of such a strategy is illustrated for maintenance problems in [10].

Before reviewing integer fractional programming we will first consider a special case of a linear fractional programming problem.

An easy linear fractional program

As we have seen, the parametric approach requires the evaluation of the parametric function F, and this is usually the most time consuming part of such a method. Therefore, it is important to analyze first the function F, before applying any of these methods. Actually, it is possible in some special cases to construct more efficient algorithms than suggested by the original approach. For instance, within the class of linear fractional programming problems the following subclass given by

$$\max_{x \in \mathcal{X}} \frac{\alpha + a^{\mathsf{T}} x}{\beta + b^{\mathsf{T}} x} \qquad (LP)$$

with $\mathcal{X} := \{x \in \mathbb{R}^m : 0 \leq x \leq 1\}$, $a \in \mathbb{R}^m$, $b \in \mathbb{R}^m_+$, $\alpha \in \mathbb{R}$ and $\beta > 0$ can be solved in polynomial time. Before proceeding with the analysis of (LP) notice that since this is a linear fractional programming problem an optimal solution will be attained at an extreme point of the polyhedron $\mathcal{X}$. Hence, $\mathcal{X}$ can be replaced by $\{0, 1\}^m$ yielding an equivalent binary fractional programming problem.

Although (LP) may seem too artificial it has several applications, like in information retrieval ([69]) and in fractional location problems as we will see in the next section. In addition, the discussion below also provides some insight on how to tackle fractional programming problems with a "nice" feasible set.

Due to the special structure of (LP) some easy simplifications can be made. Clearly, whenever b_i equals zero we can set x_i to zero if a_i is nonpositive or to one if a_i is positive. Therefore, we will consider from now on that problem (LP) is such that all the components of the vector b are positive. The associated parametric problem, illustrated in Figure 3.4, is given by

$$F(\lambda) := (\alpha - \lambda \beta) + \max_{0 \leq x \leq 1} (a - \lambda b)^{\mathsf{T}} x.$$

Taking advantage of the special form of $\mathcal{X}$, this parametric problem can be solved by inspection. Actually, an optimal solution is given by

$$x_i = \begin{cases} 1 & \text{if } a_i - \lambda b_i > 0 \text{ or } \frac{a_i}{b_i} > \lambda \\ 0 & \text{otherwise} \end{cases}$$

and its value equals

$$F(\lambda) = \alpha - \lambda \beta + \sum_{i \in I_\lambda} (a_i - \lambda b_i) \text{ where } I_\lambda := \{i \in I : \frac{a_i}{b_i} \geq \lambda\}. \qquad (3.3)$$

Notice, first that an optimal solution will have the variables such that $a_i/b_i \leq \alpha/\beta$ equal to zero. Moreover, the parametric function F is piecewise linear with a finite number of breaking points, given by a_i/b_i for $i \in I$. Clearly, the root of $F(\lambda) = 0$

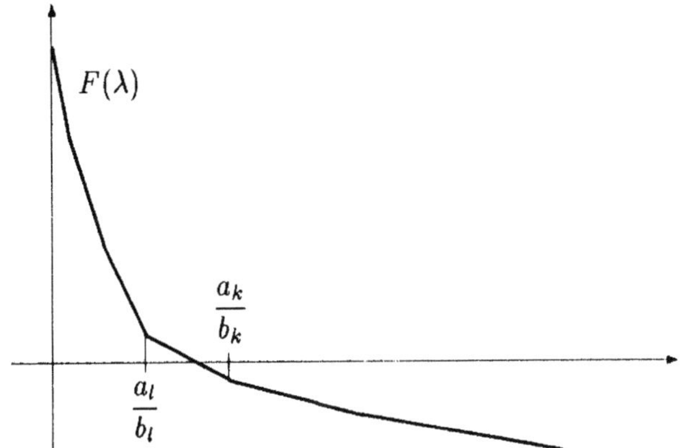

Figure 3.4: Parametric function F.

corresponds to either one of the breaking points or is a point contained in one of the pieces of the function F, see Figure 3.4. It follows that the "interesting" piece contains the breaking point

$$\frac{a_k}{b_k} := \min\left\{\frac{a_i}{b_i} : F\left(\frac{a_i}{b_i}\right) \leq 0\right\}.$$

Moreover the line containing this piece has slope $-(\beta + \sum_{i \in I_\geq} b_i)$ with $I_\geq := \{i \in I : a_i/b_i \geq a_k/b_k\}$ and hence the optimal solution value of (LP) is given by

$$\lambda_\star := \frac{\alpha + \sum_{i \in I_\geq} a_i}{\beta + \sum_{i \in I_\geq} b_i}.$$

An immediate procedure to find the "interesting" breaking point a_k/b_k is given by evaluating the function F in all its m breaking points. Since evaluating the function F at each breaking point has complexity order of $\mathcal{O}(m)$, the total complexity order of this procedure is $\mathcal{O}(m^2)$. However, if the breaking points are ordered finding a_k/b_k can be done linearly, and thus the total complexity is reduced to ordering, $\mathcal{O}(m \log m)$. Nevertheless, this computational effort can be further reduced as Hansen *et al.* [69] noticed. They introduced an ingenious procedure, described in Algorithm 3.2, that uses the well-known median finding technique.

Step 0. Let $\bar{a} := \alpha, \bar{b} := \beta$, $I_0 := \{1, \ldots, m\}$ and $I := \{i \in I_0 : p_i > \bar{a}/\bar{b}\}$;
For $i \in I_0 \setminus I$ **Do** $x_i := 0$;

Step 1. Determine $a_k/b_k := \texttt{MEDIAN}\{a_i/b_i : i \in I\}$;
Let $I_{\geq} := \{i \in I : p_i \geq a_k/b_k\}$,
$I_{\leq} := \{i \in I : p_i \leq a_k/b_k\}$ and $I_{<} := \{i \in I : p_i < a_k/b_k\}$;
Let also $\bar{a}' := \bar{a} + \sum_{i \in I_{\geq}} a_i$, $\bar{b}' := \bar{b} + \sum_{i \in I_{\geq}} b_i$ and $\lambda := \bar{a}'/\bar{b}'$;

Step 2. **If** $\lambda > a_k/b_k$ **Then GoTo** *Step* 3
Else If $I_{<} \neq \emptyset$
Then Let $a_l/b_l := \max\{a_i/b_i : i \in I_{<}\}$;
If $\lambda < a_l/b_l$ **Then GoTo** *Step* 4
Else For $i \in I_{\geq}$ **Do** $x_i := 1$;
For $i \in I_{<}$ **Do** $x_i := 0$;
Let $\lambda_{\star} := \lambda$ and **Stop**
Else For $i \in I_{\geq}$ **Do** $x_i := 1$;
Let $\lambda_{\star} := \lambda$ and **Stop**;

Step 3. **For** $i \in I_{\leq}$ **Do** $x_i := 0$;
Let $I := I \setminus I_{\leq}$ and **GoTo** *Step* 1;

Step 4. **For** $i \in I_{\geq}$ **Do** $x_i := 1$;
Let $I := I \setminus I_{\geq}$, $\bar{a} := \bar{a}'$, $\bar{b} := \bar{b}'$ and **GoTo** *Step* 1

Algorithm 3.2: Median algorithm.

Algorithm 3.2 has a complexity order of $\mathcal{O}(m)$, that can be easily explained by the following remarks. Notice first, that the initialization phase, *Step* 0, has a complexity order of $\mathcal{O}(m)$. The iterative step, a median search is performed in $\mathcal{O}(|I|)$, and or the optimal solution is found, or at least half of the remaining variables are fixed to zero or one. Therefore, the total time required by these steps is $\mathcal{O}(m) + \mathcal{O}(\frac{m}{2}) + \ldots \equiv \mathcal{O}(m)$.

Although it may not appear at first sight Algorithm 3.2 has some common points with Dinkelbach's algorithm. Notice first that at each iteration Algorithm 3.2 considers a smaller fractional problem, where the constants α and β are updated to $\bar{a}$

and $\bar{b}$.

Initiating Dinkelbach's algorithm with the median a_k/b_k used in *Step* 1 of Algorithm 3.2 requires computing $F(a_k/b_k) = \bar{a}' - a_k/b_k\bar{b}'$ to test if the current iteration point is optimal. If the current point is not optimal then Dinkelbach's algorithm takes $\lambda := \bar{a}'/\bar{b}'$ as the next iteration point. Using (3.3) it follows immediately, that checking if $\lambda > a_k/b_k$ in *Step* 2 of Algorithm 3.2 corresponds to evaluating whether $F(a_k/b_k) > 0$. Moreover by (3.3) it follows easily that if $\lambda > a_k/b_k$ then in an optimal solution all the variables x_i with $i \in I_\leq$ will be equal to zero. These reductions are performed in *Step* 3 of Algorithm 3.2, which will then pick among the remaining variables the median that will become the next iteration point. Consider now the opposite situation where $\lambda \leq a_k/b_k$, i.e. $F(a_k/b_k) \leq 0$. In Dinkelbach's algorithm we would have to compute $F(\lambda)$ to check optimality. However, as we have seen it is enough to analyze the strictly smaller breaking point adjacent to a_k/b_k, i.e. a_l/b_l, to determine if λ is an optimal point or not (see *Step* 2 of Algorithm 3.2). Clearly, if $I_<$ is an empty set then λ is the optimal solution value. Also if $I_<$ is nonempty and $\lambda \geq a_l/b_l$ then λ is the optimal point, see Figure 3.4. Notice that $\lambda \geq a_l/b_l$ implies that $F(a_l/b_l) \geq 0$ and hence $F(\lambda) = 0$. In the remaining case $\lambda < a_l/b_l < a_k/b_k$, i.e. $F(a_l/b_l) < 0$ the optimal solution point is not contained in the line segment defined by both breaking points $a_l/b_l, a_k/b_k$. It follows immediately that in an optimal solution all the variables x_i with $i \in I_\geq$ will be equal to one, see *Step* 4 of Algorithm 3.2. As the next iteration point Algorithm 3.2 considers the median of the remaining breaking points. Notice that in Dinkelbach's algorithm after computing $F(\lambda)$ the next iteration point is given by

$$\frac{\bar{a} + \sum_{i \in I_\lambda} a_i}{\bar{b} + \sum_{i \in I_\lambda} b_i}.$$

Basically the main difference between Algorithm 3.2 and Dinkelbach's algorithm lies in the choice of the new iteration point.

Algorithm 3.2 can be used for the special case of (LP) where $\alpha = 0$. Observe first, that in this case, if $a_i \leq 0$ for all $i \in I$ then the problem is trivially solved by setting all x_i equal to zero. Besides the reductions already mentioned for the general (LP) we can also set all the variables x_i to zero for which the associated a_i is negative.

3.1.2 Integer Fractional Programming

In some practical applications the variables of a fractional program are constrained to a discrete set. This class of fractional programs is usually called *integer fractional programming*. Examples of application of integer fractional programming are mostly in the field of economics, where it is important to find the biggest ratio of revenues and allocations subject to restrictions on the availability of the (indivisible) goods involved. Also combinatorial fractional programming problems, i.e. combinatorial problems where the objective function is a ratio, like *fractional cutting stock problems* [63], *minimal ratio spanning tree* [53], *fractional knapsack problems* [3, 77], *maximum mean-weight cut problem* [109, 110, 111], and *fractional location problems* [68, 95, 113] appear to be interesting from a practical point of view.

Similar to integer programming, the procedures to solve integer fractional programs are intrinsically dependent on the characteristics of the problem itself. In fact, within the class of combinatorial fractional programming problems there exist some that are quite easily solvable. Meggido [92] showed that if a problem consisting of optimizing a linear type of objective function on a certain feasible set can be solved in polynomial time then its fractional counterpart can also be solved in polynomial time, by means of an ingenious algorithm. Using this result it is possible to exhibit some combinatorial fractional programming problems that can be solved in polynomial time. Although Meggido's algorithm requires solving the usual parametric problem (P_λ) it works differently than the classical Dinkelbach algorithm. As examples, Meggido [92] mentions some applications on networks, like the minimum ratio cycles and minimum ratio spanning trees.

Recently, Radzik [109, 110, 111] has also studied several linear combinatorial fractional problems. In particular complexity results for Dinkelbach's and Meggido's algorithm are derived and existing bounds on the number of iterations are improved. One of the most interesting results is that Dinkelbach's algorithm solves a linear combinatorial fractional optimization problem in a strongly polynomial number of iterations: $\mathcal{O}(\log(mU))$ where m is the number of variables and U the maximum of the absolute values of the numerator and denominator coefficients of the variables. Some of these results are also specialized in [111] for the maximum mean-weight flow.

Also for linear combinatorial fractional programming, Karzanov [80], shows that for the subclass of problems that has as denominator the sum of all the variables, Dinkelbach's algorithm requires at most m iterations. Moreover, Hansen *et al.* [69] propose, for the even more special class of binary linear fractional programming problems[2], an algorithm running in $\mathcal{O}(m)$.

In general, also for the "difficult" integer fractional programming problems the parametric approach is used. This is mainly due to the fact that the results derived for the parametric approach are still valid for $\mathcal{X}$ a discrete set as shown in [65, 70, 129]. Moreover, in case $\mathcal{X} \subseteq I\!N^m$ the function F is piecewise linear. This implies that Dinkelbach's algorithm, and its variants, can also be directly applied to these types of problems. The key point of these applications is again evaluating the value of a parametric function, which corresponds to an integer programming problem. This strategy seems appropriate for combinatorial fractional programs, since the resulting parametric problem corresponds to a combinatorial problem of a similar nature. Examples using this approach can be found in [53, 77]. Nevertheless, as Hashizume *et al.* [70] pointed out, it is often difficult for combinatorial fractional programming problems to solve exactly the associated parametric problem. Accordingly, they propose an extension of Meggido's algorithm where the parametric problems are only solved approximately. This algorithm produces an approximate solution of the original problem with accuracy at least as good as the one of the approximated solutions of the parametric problems. An exemplification of this approach is given in [70] for the fractional knapsack problem. For the same problem, Ishii *et al.* [77] had previously proposed a solution procedure based on the Dinkelbach algorithm. This algorithm proceeds similarly as Dinkelbach's algorithm, except that the parametric problems are solved approximately. When no more improvements can be made, it is checked if the current approximate solution of the parametric problem is an optimal solution. If not, the algorithm resumes using the optimal solution of the parametric problem as the next iteration step. Recently Aneja and Kabadi [3] explored the same idea and present encouraging computational results for the fractional knapsack problem.

The parametric approach has also been used for integer linear fractional programming. Anzai [5] proposed an algorithm exploring the linearity of the problem, which

---

[2]see Section 3.1.1.

corresponds to Dinkelbach's algorithm.

Another approach to integer fractional programming is given by solving these problems directly via a branch and bound algorithm or a cutting plane algorithm. These strategies seem particularly adequate to solve (mixed) integer linear fractional programming, see [64, 65, 116]. In particular, Robillard [116] proposes for integer linear fractional programming with nondecreasing pseudo-boolean functions a branch and bound algorithm that takes advantage of the structure of the feasible set. The cutting plane approach used for linear fractional programming corresponds basically to the one used for integer linear programming, i.e. solving the linear relaxation of the problem and then progressively adding constraints that eliminate noninteger feasible points. A possible way to generate these new constraints is given by the Gomory's cuts, see [96]. Since in integer linear fractional programming the optimal solution lies at an extreme point of the feasible set, this method has been successfully applied to these types of problems, see [64, 65].

One of the approaches used for concave fractional programming consisted of transforming the original problem into a concave program. This strategy can still be used for integer fractional programming and thus the corresponding (P') problem can be solved. Notice that usually the variables of the transformed problem will now be restricted to a discrete set formed usually by noninteger values. For the linear case, Granot and Granot [64] avoid this obstacle by considering another transformed problem based on the same transformation and on the "linearity" of the involved functions and feasible set.

3.2 Fractional Location Models

As discussed at the beginning of this chapter there exist some practical situations which can be modeled as fractional programming problems. In particular, within discrete location theory it is reasonable and desirable in some situations to maximize the profitability index instead of the total net profit, as mentioned by Revelle and Laporte [113] in their survey on new directions in facility location. Therefore, after briefly surveying the few existing references we will start by discussing solution procedures for the model proposed by Revelle and Laporte [113]. This analysis will motivate the study of some variations of this model, including the usual uncapacitated facility location problem with a ratio as an objective function. Among these apparently difficult models we will identify those problems that can be solved in polynomial time with respect to the problem dimension, i.e. number of clients and number of facilities. The approach followed to analyze the 1-level type of location models will be used to study the 2-level models discussed in the previous chapter. These results, based on Barros *et al.* [13], are here discussed in detail.

3.2.1   1-level Fractional Location Problems

An example of the recent interest in fractional location models is given by the practical situation of locating and sizing offshore oil platforms. Initially, Hansen *et al.* [67] modeled this problem as a *multicapacitated facility location problem* where the investment costs have to be minimized. However, Rosing [119] suggested that for this practical situation it would be preferable to minimize the cost/production ratio. Subsequently, Hansen *et al.* [68] considered in the model this ratio as the objective function and proposed to solve this combinatorial fractional programming problem using Dinkelbach's algorithm.

Another example of a fractional location problem is given by Myung and Tcha [95] who propose to study a generalization of the uncapacitated facility location problem. In this model the profitability index is maximized subject to the usual constraints of an uncapacitated facility location problem plus an additional constraint to guarantee a minimum net profit π. Using the same notation and terminology as adopted in Chapter 2, this problem can be formalized as follows

$$\max \quad \frac{\sum_{i \in I} \sum_{j \in J} c_{ij} x_{ij}}{\sum_{j \in J} f_j y_j}$$

$$\text{s.t.:} \sum_{i \in I} \sum_{j \in J} c_{ij} x_{ij} - \sum_{j \in J} f_j y_j \geq \pi$$

$$\sum_{j \in J} x_{ij} = 1 \qquad \forall i \in I$$

$$x_{ij} \leq y_j \qquad \forall i \in I, j \in J$$

$$x_{ij} \geq 0, y_j \in \{0, 1\} \qquad \forall i \in I, j \in J$$

where it is assumed that $f_j > 0$, $j \in J$ and $\pi > 0$. In order to solve this combinatorial fractional program Myung and Tcha [95] use Dinkelbach's algorithm. Their choice is justified by the fact that the associated parametric problem corresponds to a standard uncapacitated facility location problem with the additional constraint on the minimal amount of net profit. This characteristic enables the derivation of several properties and results which simplify the solution of the successive parametric problems.

Finally, Revelle and Laporte [113] discuss the situation where the location of facilities has to be decided according to potential clients. This model is suitable for those practical situations, where no pre-established contracts with clients exist and therefore, there is not any obligation of satisfying the demand of the clients. Revelle and Laporte [113] mention that in this case it is important to maximize the ratio between the profits and the associated investments. The profits associated with serving client i via facility j are determined by a function of the production costs at facility j, the demand d_i of client i, the selling price of client i and the transportation costs between client i and facility j. On the other hand, the investments consider not only the usual fixed costs f_j of opening the facilities but also the so-called expansion costs e_j of manufacturing. The expansion costs at facility j associated with client i can be viewed as the costs of purchasing the machinery necessary to produce the demand d_i at site j, i.e. $\sum_{i \in I} \sum_{j \in J} d_i e_j x_{ij}$. The above model was formulated by Revelle and Laporte [113] as follows

$$\max \frac{\sum_{i \in I} \sum_{j \in J} c_{ij} x_{ij}}{\sum_{j \in J} f_j y_j + \sum_{i \in I} \sum_{j \in J} d_i e_j x_{ij}} \qquad (RL)$$

$$\text{s.t.:} \quad \sum_{j \in J} x_{ij} \leq 1 \qquad \forall i \in I \qquad (3.4)$$

$$x_{ij} \leq y_j \qquad \forall i \in I, j \in J \qquad (3.5)$$

$$x_{ij} \geq 0, y_j \in \{0,1\} \quad \forall i \in I, j \in J \qquad (3.6)$$

without any further assumptions. Notice that the above formulation is not well defined since the objective function is not defined at $(0, \ldots, 0)$ which is a feasible solution. However, from an investor's point of view such a solution is not interesting since it corresponds to not investing at all. Moreover, the profitability index is used as an economical criterion whenever there are diverse investment options. Hence, it is essential to check *a priori* if not investing is the only optimal solution.

Before continuing this analysis we will need to assume that the fixed costs f_j are *positive* and that the demand d_i of client i as well as the expansion costs e_j of a facility are *nonnegative*. Clearly, if it is decided that an investment is made, one or more facilities will be opened, and thus the optimal solution of (RL) will not be changed by adding the constraint

$$\sum_{j \in J} y_j \geq 1. \tag{3.7}$$

Problem (RL) with this additional constraint corresponds to a properly defined integer fractional programming problem. Clearly, if all the coefficients c_{ij} are nonpositive, an optimal solution would be to serve none of the clients, i.e. $x_{ij} = 0$ for every $i \in I$ and $j \in J$, and hence our initial decision to invest in this location project would be irrational. Therefore, we will assume from now on that serving at least one client via some facility gives a positive profit, i.e. one of the coefficients c_{ij} is positive.

The combinatorial structure of (RL) immediately suggests the use of a parametric approach, although the associated parametric problem is an $\mathcal{NP}$-hard problem. Notice that without the inclusion of the extra constraint (3.7) the associated parametric problem is given by

$$F(\lambda) := \max \sum_{i \in I} \sum_{j \in J} c_{ij} x_{ij} - \lambda \left(\sum_{j \in J} f_j y_j + \sum_{i \in I} \sum_{j \in J} d_i e_j x_{ij} \right) \tag{RL_λ}$$

$$\text{s.t.:} \quad \sum_{j \in J} x_{ij} \leq 1 \qquad \forall i \in I$$

$$x_{ij} \leq y_j \qquad \forall i \in I, j \in J$$

$$x_{ij} \geq 0, y_j \in \{0, 1\} \quad \forall i \in I, j \in J.$$

The main properties of the above parametric problem are summarized in the next lemma.

Lemma 3.2.1

If at least one of the coefficients c_{ij} is positive then the parametric function F is continuous, decreasing and $F(\lambda) \geq 0$ for all $\lambda \in \mathbb{R}$. Moreover, the optimal value of

(RL_λ) is strictly positive if and only if $\lambda < \vartheta(RL)$, and equal to zero if and only if $\lambda \geq \vartheta(RL)$. Furthermore, for $\lambda = \vartheta(RL)$ the optimal solution set of (RL_λ) with the null vector excluded equals the optimal solution set of (RL).

Proof: Clearly, the function F is continuous and decreasing. Since $(0, \ldots, 0)$ is a feasible solution of (RL_λ) it follows immediately that $F(\lambda) \geq 0$ for every $\lambda \in \mathbb{R}$.

In order to prove the second part observe first that for $F(\lambda) > 0$ all optimal solutions of (RL_λ) are different from the null vector. Such an optimal solution of (RL_λ) will generate a lower bound on $\vartheta(RL)$, and this lower bound, using $F(\lambda) > 0$, is greater than λ. Hence if $F(\lambda) > 0$ then $\lambda < \vartheta(RL)$. Moreover, if $\lambda < \vartheta(RL)$ we obtain since $\vartheta(RL) > 0$ that any optimal solution $(y^\star, x^\star)$ of (RL) is different from the null vector and so

$$\sum_{j \in J} f_j y_j^\star + \sum_{i \in I} \sum_{j \in J} d_i e_j x_{ij}^\star > 0.$$

This implies that

$$F(\lambda) \geq \sum_{i \in I} \sum_{j \in J} c_{ij} x_{ij}^\star - \lambda \left(\sum_{j \in J} f_j y_j^\star + \sum_{i \in I} \sum_{j \in J} d_i e_j x_{ij}^\star \right)$$

$$> \sum_{i \in I} \sum_{j \in J} c_{ij} x_{ij}^\star - \vartheta(RL) \left(\sum_{j \in J} f_j y_j^\star + \sum_{i \in I} \sum_{j \in J} d_i e_j x_{ij}^\star \right) = 0$$

and hence the second result is proved.

To verify the third result it is sufficient to show that $F(\lambda) = 0$ for every $\lambda \geq \vartheta(RL)$. From the previous results we have that $F(\lambda) \leq 0$ for every $\lambda \geq \vartheta(RL)$ and since $F(\lambda) \geq 0$ for every $\lambda \in \mathbb{R}$ it follows that $F(\lambda) = 0$ for $\lambda \geq \vartheta(RL)$.

Finally, the last result is easy to verify and so its proof is omitted. $\qquad \square$

The above lemma provides a geometrical interpretation for the parametric function associated with (RL_λ), see Figure 3.5.

Lemma 3.2.1 also suggests applying Dinkelbach's algorithm to the original problem (RL) starting with a lower bound on the optimal value (given by any feasible solution different from the null vector). Observe that this approach is similar to the one proposed by Barros and Frenk [12] to another class of fractional programs.

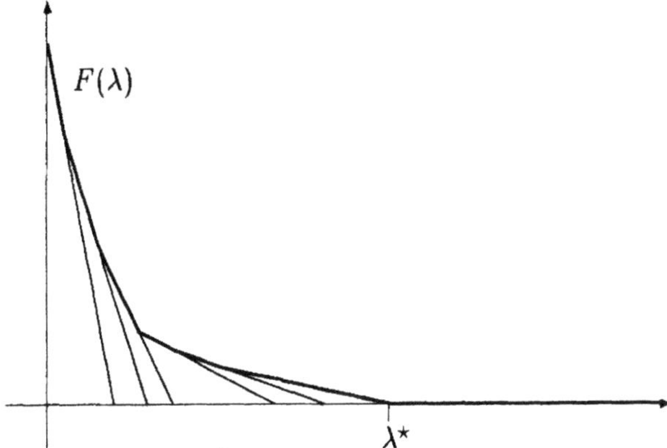

Figure 3.5: Parametric function associated with (RL).

The approach described above can be seen as an application of basic tools in fractional programming to this combinatorial fractional program. However, due to the special structure of (RL) it is possible to directly solve this problem using an algorithm with a running time dependent on the number of facilities p and clients m. Observe first that (RL_λ) corresponds to a special uncapacitated facility location problem where the usual assignment constraints $\sum_{j \in J} x_{ij} = 1$ are replaced by (3.4)

$$\max \sum_{i \in I} \sum_{j \in J} c_{ij} x_{ij} - \sum_{j \in J} f_j y_j \qquad (RUFLP)$$

$$\text{s.t.:} \quad \sum_{j \in J} x_{ij} \leq 1 \qquad \forall i \in I$$

$$x_{ij} \leq y_j \qquad \forall i \in I, j \in J$$

$$x_{ij} \geq 0, y_j \in \{0, 1\} \quad \forall i \in I, j \in J.$$

From Lemma 3.2.1 it follows that for λ equal to $\vartheta(RL)$ the value of the associated parametric problem is zero. This suggests to investigate the above type of location problems which have zero as the optimal value. Observe that in the next lemma no assumptions on the coefficients c_{ij} and f_j, $i \in I, j \in J$ are required.

Lemma 3.2.2

If the optimal value of (RUFLP) equals zero, then either the null vector is the only optimal solution to this problem or there exists an optimal solution with only one facility open.

Proof: Clearly $(0, \ldots, 0)$ is an optimal solution of $(RUFLP)$. Assume now that this is not the only optimal solution and let $(\boldsymbol{y}, \boldsymbol{x})$ be an optimal solution of $(RUFLP)$, with the set of open facilities given by $J^\star \neq \emptyset$. By assumption we have that

$$0 = \vartheta(RUFLP) = \sum_{j \in J^\star} \left(\sum_{i \in I} c_{ij} x_{ij} - f_j \right). \tag{3.8}$$

Let $j^\star \in J^\star$. If $\sum_{i \in I} c_{ij^\star} x_{ij^\star} - f_{j^\star} > 0$ then the solution $(\boldsymbol{y}', \boldsymbol{x}')$ given by

$$y'_j = \begin{cases} 1 & \text{if } j = j^\star \\ 0 & \text{otherwise} \end{cases} \quad \text{and } x'_{ij} = \begin{cases} x_{ij^\star} & \text{for all } i \in I, j = j^\star \\ 0 & \text{otherwise} \end{cases} \tag{3.9}$$

is feasible with a positive objective value and this contradicts $\vartheta(RUFLP) = 0$. Hence, for every $j^\star \in J^\star$ it follows that $\sum_{i \in I} c_{ij^\star} x_{ij^\star} - f_{j^\star} \leq 0$ and due to (3.8) we must have that $\sum_{i \in I} c_{ij^\star} x_{ij^\star} - f_{j^\star} = 0$ for every $j^\star \in J^\star$. Therefore, for any $j^\star \in J^\star$, the solution given by (3.9) solves $(RUFLP)$. $\square$

We can now characterize an optimal solution of (RL).

Proposition 3.2.1

If at least one of the coefficients c_{ij} is positive there exists an optimal solution of (RL) with only one facility open. Moreover,

$$\vartheta(RL) = \max_{j \in J} \left\{ \max \left\{ \frac{\sum_{i \in I} c_{ij} x_{ij}}{f_j + \sum_{i \in I} d_i e_j x_{ij}} : 0 \leq x_{ij} \leq 1, \text{ for all } i \in I \right\} \right\}. \tag{3.10}$$

Proof: From Lemma 3.2.1 it follows that the associated parametric problem has optimal value equal to zero for $\lambda = \vartheta(RL) > 0$ and there exists a nonzero optimal solution of this parametric problem. Hence, by Lemma 3.2.2 we can find an optimal solution of (RL_λ) with only one facility open. Applying again Lemma 3.2.1 yields the existence of an optimal solution of (RL) with only one facility open, and hence the first result is proved. Using this result (3.10) follows trivially. $\square$

Observe that in order to determine the optimal value of (RL) we must solve (3.10), which corresponds to solving p simple allocation problems. Moreover, each of these allocation problems corresponds to a linear fractional programming problem of the form

$$\max_{0 \le x_i \le 1} \frac{\sum_{i \in I} a_i x_i}{\beta + \sum_{i \in I} b_i x_i}$$

which can be solved easily by Algorithm 3.2 described in Section 3.1.1. Since this algorithm has a complexity order of $\mathcal{O}(m)$ the next result is easy to derive.

Theorem 3.2.1

If at least one coefficient c_{ij} is positive, the fractional location problem (RL) can be solved in polynomial time with a complexity order of at most $\mathcal{O}(p\,m)$.

The model proposed by Revelle and Laporte [113] assumes the existence of expansion costs. Clearly, if these costs are zero the above results are still valid. Moreover, in this case the allocation problem (3.10) reduces to a trivial linear programming problem.

It is now interesting to analyze for which generalizations of (RL) the above result is still valid. Therefore, we will assume that the investments also involve an additional initial fixed investment $f_0 > 0$. In this case the objective function of (RL) is replaced by

$$\max \frac{\sum_{i \in I} \sum_{j \in J} c_{ij} x_{ij}}{f_0 + \sum_{j \in J} f_j y_j + \sum_{i \in I} \sum_{j \in J} d_i e_j x_{ij}}. \qquad (FRUFLPf)$$

As the next example shows the previous results are no longer valid for this model.

Example 3.2.1

Let $I = \{1, 2\}$ denote the set of clients, $J = \{1, 2\}$ the set of facilities and consider the following coefficients

c_{ij}	1	2		$d_i\, e_j$	1	2
1	20	5		1	5	5
2	5	20		2	10	10

Let also the fixed costs of both facilities be equal to 5 and the initial investment f_0 be equal to 10.

Clearly, if only one facility is open the maximum profitability index is equal to 1, while if both facilities are open the maximum profitability index equals $\frac{8}{7}$.

3.2. Fractional Location Models

An obvious variation of the model proposed by Revelle and Laporte [113] is given by assuming that the demand of the clients needs to be satisfied. In this case we have a standard uncapacitated facility location problem where the profitability index has to be maximized. We will start this analysis by considering that there are no expansion costs, i.e. $e_j = 0$, and that $f_j > 0$ for every $j \in J$.

$$\max \frac{\sum_{i \in I} \sum_{j \in J} c_{ij} x_{ij}}{\sum_{j \in J} f_j y_j} \qquad (FUFLP)$$

$$\text{s.t.:} \quad \sum_{j \in J} x_{ij} = 1 \qquad \forall i \in I \qquad (3.11)$$

$$x_{ij} \leq y_j \qquad \forall i \in I, j \in J$$

$$x_{ij} \geq 0, y_j \in \{0, 1\} \quad \forall i \in I, j \in J.$$

Observe that if there would exist a facility j such that $f_j = 0$ and $c_{ij} = 0$ for all $i \in I$ then $(FUFLP)$ could be simplified by removing this facility and replacing constraint (3.11) by (3.4). This simplified problem reduces to the optimization problem (RL) with no expansion costs, and hence the results previously derived are valid for this particular situation.

Due to the combinatorial nature of the above integer fractional programming problem, it seems quite reasonable to apply a parametric approach to solve $(FUFLP)$. Therefore, consider the associated parametric problem,

$$\max \sum_{i \in I} \sum_{j \in J} c_{ij} x_{ij} - \lambda \sum_{j \in J} f_j y_j \qquad (FUFLP_\lambda)$$

$$\text{s.t.:} \quad \sum_{j \in J} x_{ij} = 1 \qquad \forall i \in I$$

$$x_{ij} \leq y_j \qquad \forall i \in I, j \in J$$

$$x_{ij} \geq 0, y_j \in \{0, 1\} \quad \forall i \in I, j \in J$$

which corresponds to an uncapacitated facility location problem. A similar result, as obtained in Lemma 3.2.2 can be derived. However, due to constraints (3.11) we need some additional assumptions on the data.

Lemma 3.2.3

If $c_{ij} \geq 0$ for all $i \in I$, $j \in J$, $f_j \geq 0$ for all $j \in J$ and the optimal value of

(UFLP) equals zero, then there exists an optimal solution to this problem with only one facility open.

Proof: Let (y, x) be a nonzero optimal solution of the problem $(UFLP)$, with the nonempty set of open facilities given by J^*. By assumption we have that

$$0 = \vartheta(UFLP) = \sum_{j \in J^*} \left(\sum_{i \in I} c_{ij} x_{ij} - f_j \right). \qquad (3.12)$$

Let $j^* \in J^*$. If $\sum_{i \in I} c_{ij^*} x_{ij^*} - f_{j^*} > 0$ then by the nonnegativity assumption on c_{ij} we also have that

$$\sum_{i \in I} c_{ij^*} - f_{j^*} \geq \sum_{i \in I} c_{ij^*} x_{ij^*} - f_{j^*} > 0.$$

This implies that the solution (y', x') given by

$$y'_j = \begin{cases} 1 & \text{if } j = j^* \\ 0 & \text{otherwise} \end{cases} \quad \text{and} \quad x'_{ij} = \begin{cases} 1 & \text{for all } i \in I \text{ and } j = j^* \\ 0 & \text{otherwise} \end{cases} \qquad (3.13)$$

is feasible with a positive objective value. Since $\vartheta(UFLP) = 0$, this contradicts the optimality of (y, x) and so for $j^* \in J^*$ we must have by (3.12) that $\sum_{i \in I} c_{ij^*} x_{ij^*} - f_{j^*} = 0$. This implies that

$$\sum_{i \in I} c_{ij^*} - f_{j^*} \geq \sum_{i \in I} c_{ij^*} x_{ij^*} - f_{j^*} = 0$$

and since the solution given by (3.13) is feasible it follows that $\sum_{i \in I} c_{ij^*} - f_{j^*} \leq \vartheta(UFLP) = 0$. Hence, $\sum_{i \in I} c_{ij^*} - f_{j^*} = 0$ and so (3.13) solves $(UFLP)$ which concludes the proof. $\qquad \Box$

Using the above lemma it is possible to characterize an optimal solution of $(FUFLP)$.

Proposition 3.2.2

If $c_{ij} \geq 0$ for all $i \in I$, $j \in J$, $f_j > 0$ for all $j \in J$ then an optimal solution of $(FUFLP)$ corresponds to open only one facility. Moreover,

$$\vartheta(FUFLP) = \max_{j \in J} \frac{\sum_{i \in I} c_{ij}}{f_j}.$$

Proof: Notice first that by Lemma 3.1.1 it follows that the objective value of the parametric problem $(UFLP_{\lambda_\star})$ equals zero for $\lambda_\star = \vartheta(FUFLP)$. Hence, by Lemma 3.2.3 (with f_j replaced by $\lambda_\star f_j$) we can find an optimal solution of $(UFLP_{\lambda_\star})$ with only one facility open and so again by Lemma 3.1.1 the first part is proved. The second part follows now easily. $\square$

Using the above proposition we have the next result.

Theorem 3.2.2

If $c_{ij} \geq 0$ for all $i \in I$, $j \in J$, $f_j > 0$ for all $j \in J$ then $(FUFLP)$ can be solved linearly in $\mathcal{O}(p\,m)$ time.

Whenever the assumption on the profitability of the clients, i.e. $c_{ij} \geq 0$ for all $i \in I$, $j \in J$ cannot be ensured then $(FUFLP)$ may have only optimal solutions with more than one facility open. This can be easily shown by considering in Example 3.2.1 $c_{12} = c_{21} = -5$ and removing the initial investment as well as the expansion costs. In this case, opening only one facility yields a profitability index of 3 while opening the two facilities yields a bigger profitability index of 4.

There exist some additional modifications to the pure fractional uncapacitated location model $(FUFLP)$ which transform this problem into one for which Proposition 3.2.2 does not hold. For instance, whenever it is assumed that the investment is composed not only of the fixed costs of opening facilities, but also of an additional initial fixed investment $f_0 > 0$. The model $(FUFLP)$ will then consider the following objective function

$$\max \frac{\sum_{i \in I} \sum_{j \in J} c_{ij} x_{ij}}{f_0 + \sum_{j \in J} f_j y_j}. \qquad (FUFLPf)$$

In this case it is not possible to guarantee that there exists one optimal solution of $(FUFLPf)$ which has only one facility open. In fact, Example 3.2.1 without expansion costs shows that with only facility open the maximum profitability index is $5/3$, while with both open we have 2. Also by including in $(FUFLP)$ some expansion costs $e_j > 0$ as in the model of Revelle and Laporte [113], the above results are no longer valid as Example 3.2.1 without initial fixed costs shows. In this case, the biggest profitability index is attained when both facilities are open.

Observe that for similar reasons the model proposed by Myung and Tcha [95], cannot be solved by inspection (unless $\pi = 0$).

In these "difficult" cases, parametric approaches like the ones in [3, 77] appear to be suitable solution techniques.

3.2.2 2-level Fractional Location Problems

So far we have only considered 1-level fractional location problems. We will now try to extend the results established for the 2-level models discussed in the previous chapter.

In the general model proposed in Section 2.4 the objective function corresponds to maximizing the total net profit. Considering now as the objective function the ratio between the profits and the total sum of fixed costs

$$\frac{\sum_{i \in I} \sum_{j \in J} \sum_{k \in K} c_{ijk} x_{ijk}}{\sum_{j \in J} f_j y_j + \sum_{k \in K} g_k z_k + \sum_{j \in J} \sum_{k \in K} F_{jk} t_{jk}},$$

we have a fractional programming problem. Unfortunately, it is not possible to adapt Lemma 3.2.3 to this case, due to the simultaneous presence of the individual fixed costs f_j of the facilities and g_k of the depots. This suggests to consider $(2ELP)$, i.e. the model proposed by Gao and Robinson [59], see Section 2.2. The corresponding fractional model is given by

$$\max \frac{\sum_{i \in I} \sum_{j \in J} \sum_{k \in K} c_{ijk} x_{ijk}}{\sum_{j \in J} f_j y_j + \sum_{j \in J} \sum_{k \in K} F_{jk} t_{jk}} \qquad (F2ELP)$$

$$\text{s.t.:} \sum_{j \in J} \sum_{k \in K} x_{ijk} = 1 \quad \forall i \in I$$

$$x_{ijk} \leq t_{jk} \qquad \forall i \in I, j \in J, k \in K$$

$$t_{jk} \leq y_j \qquad \forall j \in J, k \in K$$

$$y_j, t_{jk} \in \{0, 1\} \quad \forall j \in J, k \in K$$

$$x_{ijk} \geq 0 \qquad \forall i \in I, j \in J, k \in K.$$

We will also assume that $f_j > 0$, $j \in J$ and $F_{jk} > 0$, $j \in J, k \in K$. A similar result, to that obtained in Lemma 3.2.3 can be derived.

Lemma 3.2.4

If $c_{ijk} \geq 0$ for all $i \in I$, $j \in J$, $k \in K$, and the optimal objective value of the (2ELP) equals zero, then there exists an optimal solution to this problem with only one facility open.

Proof: Let $(\boldsymbol{t}, \boldsymbol{y}, \boldsymbol{x})$ be an optimal solution of the (2ELP). Notice that this solution is different from the null vector and denote the set of operating pairs by $T^\star$. Let also $J^\star$ define the set of open facilities and K_j represent the set of depots operating with facility $j \in J^\star$. By assumption we have that

$$0 = \vartheta(2ELP) = \sum_{j \in J^\star} \left(\sum_{k \in K_j} \left(\sum_{i \in I} c_{ijk} x_{ijk} - F_{jk} \right) - f_j \right).$$

Let $j^\star \in J^\star$. If

$$\sum_{k \in K_{j^\star}} \left(\sum_{i \in I} c_{ij^\star k} x_{ij^\star k} - F_{j^\star k} \right) - f_{j^\star} > 0$$

then by the nonnegativity assumption we also have that

$$\sum_{i \in I} \max_{k \in K_{j^\star}} c_{ij^\star k} - \sum_{k \in K_{j^\star}} F_{j^\star k} - f_{j^\star} \geq \sum_{k \in K_{j^\star}} \left(\sum_{i \in I} c_{ij^\star k} x_{ij^\star k} - F_{j^\star k} \right) - f_{j^\star} > 0.$$

This implies that the solution $(\boldsymbol{t}', \boldsymbol{y}', \boldsymbol{x}')$ given by

$$t'_{jk} = \begin{cases} 1 & \text{if } j = j^\star, k \in K_{j^\star} \\ 0 & \text{otherwise} \end{cases} , \quad y'_j = \begin{cases} 1 & \text{if } j = j^\star \\ 0 & \text{otherwise} \end{cases} , \qquad (3.14)$$

$$x'_{ijk} = \begin{cases} 1 & \text{if } j = j^\star, k = \arg \max_{\ell \in K_{j^\star}} c_{ij^\star \ell} \\ 0 & \text{otherwise} \end{cases}$$

is feasible with a positive objective value. Since $\vartheta(2ELP) = 0$ this contradicts the optimality of $(\boldsymbol{t}, \boldsymbol{y}, \boldsymbol{x})$ and so for $j^\star \in J^\star$ we must have that

$$\sum_{k \in K_{j^\star}} \left(\sum_{i \in I} c_{ij^\star k} x_{ij^\star k} - F_{j^\star k} \right) - f_{j^\star} = 0.$$

As in the proof of Lemma 3.2.3 it follows, for any $j^\star \in J^\star$, that the solution given by (3.14) solves (2ELP) which concludes the proof. $\square$

Using the above lemma it is possible by a similar reasoning as in Proposition 3.2.2 to partially characterize an optimal solution of $(F2ELP)$.

Proposition 3.2.3

If $c_{ijk} \geq 0$ for all $i \in I$, $j \in J$, $k \in K$ and f_j, $F_{jk} > 0$ for all $j \in J$, $k \in K$ then there exists an optimal solution of $(F2ELP)$ with only one facility open.

The above proposition shows that instead of solving $(F2ELP)$ directly we can decompose the original problem into p fractional location problems of the form

$$\max \frac{\sum_{i \in I} \sum_{k \in K} c_{ijk} x_{ijk}}{f_j + \sum_{k \in K} F_{jk} t_{jk}} \qquad (F2ELP_j)$$

$$\text{s.t.:} \sum_{k \in K} x_{ijk} = 1 \quad \forall i \in I$$

$$x_{ijk} \leq t_{jk} \quad \forall i \in I, k \in K$$

$$t_{jk} \in \{0, 1\} \quad \forall k \in K$$

$$x_{ijk} \geq 0 \quad \forall i \in I, k \in K.$$

Observe that the above problem corresponds to $(FUFLPf)$, since now variables t_{jk} correspond to the "facilities" to be located. Therefore, we cannot immediately exhibit an optimal solution. However, if $f_j = 0$ then this problem reduces to an instance of $(FUFLP)$ and hence, by Proposition 3.2.2, it can be easily solved.

If pre-existing contracts with the clients do not exist, then the usual assignment constraints $\sum_{j \in J} \sum_{k \in K} x_{ijk} = 1$ can be replaced by

$$\sum_{j \in J} \sum_{k \in K} x_{ijk} \leq 1 \quad \forall i \in I \qquad (3.15)$$

yielding the following 2-level location problem

$$\max \frac{\sum_{i \in I} \sum_{j \in J} \sum_{k \in K} c_{ijk} x_{ijk}}{\sum_{j \in J} f_j y_j + \sum_{j \in J} \sum_{k \in K} F_{jk} t_{jk}} \qquad (FR2ELP)$$

$$\text{s.t.:} \sum_{j \in J} \sum_{k \in K} x_{ijk} \leq 1 \quad \forall i \in I$$

$$x_{ijk} \leq t_{jk} \quad \forall i \in I, j \in J, k \in K$$

$$t_{jk} \leq y_j \qquad \forall j \in J, k \in K$$

$$y_j, t_{jk} \in \{0,1\} \qquad \forall j \in J, k \in K$$

$$x_{ijk} \geq 0 \qquad \forall i \in I, j \in J, k \in K.$$

Similar to the 1-level fractional location problem (RL), the above fractional problem is no longer well defined. However, the reasoning used to discuss the 1-level case can be adapted as follows. Clearly, if it is decided that an investment is made at least one pair facility j and depot k is opened and hence the optimal solution of $(FR2ELP)$ does not change by adding the following constraint

$$\sum_{j \in J} \sum_{k \in K} t_{jk} \geq 1.$$

Observe that now we have a well defined fractional programming problem. Such an investment is only logical if at least one client yields profit, and thus we also consider that at least one $c_{ijk} > 0$. A similar analysis as done for Proposition 3.2.1, by means of Lemma 3.2.2, leads to the following result.

Proposition 3.2.4

If there exists a coefficient c_{ijk} positive and f_j, $F_{jk} > 0$ for all $j \in J$, $k \in K$ then there exists an optimal solution of $(FR2ELP)$, with only one facility open.

Notice that the above result implies that $(FR2ELP)$ can be decomposed into p 1-level fractional location problems of the form

$$\max \frac{\sum_{i \in I} \sum_{k \in K} c_{ijk} x_{ijk}}{f_j + \sum_{k \in K} F_{jk} t_{jk}} \qquad (FR2ELPj)$$

$$\text{s.t.:} \sum_{k \in K} x_{ijk} \leq 1 \quad \forall i \in I$$

$$x_{ijk} \leq t_{jk} \qquad \forall i \in I, k \in K$$

$$t_{jk} \in \{0,1\} \qquad \forall k \in K$$

$$x_{ijk} \geq 0 \qquad \forall i \in I, k \in K.$$

Observe that the above problem is now a $(FRUFLPf)$ and hence we cannot immediately exhibit an optimal solution. However, if $f_j = 0$ then this problem reduces to (RL) and hence it can be easily solved.

Unfortunately, similar results cannot also be derived for the 2-level uncapacitated facility location problem discussed in the previous chapter. Apparently the fractional models associated with both the general model and the pure 2-level location problems are more "intricate" than $(F2ELP)$. It is interesting to notice that the computational experience performed in the first chapter also showed that from the three variants the $(2ELP)$ appears to be the easiest.

3.3   Conclusions

Within location theory, models with a ratio as objective function have not been explored as much as the usual linear types of objectives functions. This can be explained by the fact that fractional programming has always been regarded as a specialized field within nonlinear programming. In addition, integer fractional programming appears to be, in the perspective of nonlinear programming, a rather "unsmooth" field. The above two reasons probably justify the little interest showed so far on fractional location models. The recent references on this subject seem to demonstrate both the importance and utility of such models.

In this chapter besides reviewing briefly fractional programming we analyzed some variants of the uncapacitated facility location problem with a ratio as objective function. Using basic concepts and results of fractional programming it was possible to identify among these 1-level fractional location problems, the ones which can be solved in polynomial time. The solution procedure encountered is based on the fact that for these problems, (RL) and $(FUFLP)$, there exists an optimal solution with only one facility open. However, the above result requires that in $(FUFLP)$ all the profits should be nonnegative. As expected, the 2-level fractional location problems revealed to be more "intricate" than their 1-level counterpart. In fact, only for the fractional 2-echelon uncapacitated facility location problem it was possible to characterize partially an optimal solution, under the same assumption as for $(FUFLP)$. In this case it was shown that there exists an optimal solution of this problem with one facility open. However, it is left to determine which depots are operating with the open facility. This yields that solving the 2-echelon fractional location problem corresponds to decomposing this problem into p 1-level fractional location problems. These results show that there exist some classes of location problems inherently difficult, which have easy fractional counterparts.

Finally, we would like to point out that the theoretical results obtained in this chapter are easily obtained by combining techniques of the two apparent distinct fields of fractional programming (nonlinear programming) and combinatorics. Moreover, this combined approach permits to identify among apparently difficult problems the ones which can be solved easily.

Four

Generalized Fractional Programming

In the previous chapter we considered some location models with a ratio as objective function. This discussion showed the importance of having a good knowledge of fractional programming when solving a fractional location model. As we shall see in Section 4.1.2 there exist also models in location analysis which are actually generalized fractional programs. Therefore, we will devote this chapter to *generalized fractional programming*, where the goal is to minimize the largest of several ratios of functions. One of the earliest examples of generalized fractional programming problem corresponds to Von Neumann's model of an expanding economy, see [100]. Other applications can be found in economic equilibrium problems, management applications of goal programming and multi-objective programming involving ratios of functions, and rational approximation in numerical analysis, see [42, 124]. Let $\mathcal{X} \subset \mathbb{R}^m$ be a nonempty set and $f_j, g_j : \mathcal{C} \longrightarrow \mathbb{R}$, $j \in J := \{1, \ldots, n\}$, $n \geq 1$, be a class of continuous functions where $\mathcal{C}$ is an open set containing $\mathcal{X}$. Assuming that $g_j(\boldsymbol{x}) > 0$ for every $\boldsymbol{x} \in \mathcal{X}$ and $j \in J$, a generalized fractional program is defined as follows

$$\inf_{\boldsymbol{x} \in \mathcal{X}} \max_{j \in J} \frac{f_j(\boldsymbol{x})}{g_j(\boldsymbol{x})}. \tag{P}$$

Observe that by definition the problem (P) is solvable if there exists some $\boldsymbol{x} \in \mathcal{X}$ such that $\vartheta(P)$ equals $\psi(\boldsymbol{x})$, with $\psi : \mathcal{C} \longrightarrow \overline{\mathbb{R}}$ given by

$$\psi(\boldsymbol{x}) := \max_{j \in J} \frac{f_j(\boldsymbol{x})}{g_j(\boldsymbol{x})}. \tag{4.1}$$

Clearly, for $n = 1$ problem (P) reduces to a single-ratio fractional programming problem, see Section 3.1.

Similar to fractional programming, an important class of generalized fractional programs is given by *convex generalized fractional programs*. This class of problems is such that the feasible set $\mathcal{X}$ is convex and given by $\mathcal{X} := \{x \in \mathcal{S} : h(x) \leq 0\}$, with $h : \mathbb{R}^m \longrightarrow \mathbb{R}^r$ a convex vector-valued function and $\mathcal{S} \subseteq \mathcal{C}$ a convex set. Additionally, the functions $f_j, g_j : \mathcal{C} \longrightarrow \mathbb{R}$ are respectively convex and concave on $\mathcal{S}$ and g_j are positive on $\mathcal{X}$ for all $j \in J$. Moreover, either f_j is a nonnegative function on $\mathcal{X}$ or g_j is affine. For this class of functions it follows that the function ψ, being the maximum of semistrictly quasiconvex functions, is again a semistricly quasiconvex function on $\mathcal{X}$ and so a local minimum of (P) is a global minimum, see [6]. These properties make this class of generalized fractional programs more attractive and hence most of the research is focused on convex generalized fractional programs. This important class also includes *generalized linear fractional programs*, which as the name suggests consider only ratios of affine functions and $\mathcal{X}$ a convex polyhedron.

Contrary to fractional programming, the literature on generalized fractional programming is not so abundant in different solution procedures. Most of the solution procedures available are parametric type of approaches based on the pioneering paper of Crouzeix *et al.* [43]. In this paper, the well-known Dinkelbach procedure for classical fractional programming is extended to the multi-ratio case. Variants and extensions of this approach can be found in [41].

In the special case of convex generalized fractional programming problems the objective function is subdifferentiable and Boncompte and Martínez-Legaz [28] take advantage of this characteristic to apply a cutting plane algorithm. Recently Nesterov and Nemirovsky [135] have proposed a Karmakar-like interior point algorithm for generalized linear fractional programming based on the concept of self-concordance barriers. Also, Freund and Jarre [55] present an interior point algorithm for the class of generalized fractional programming formed by ratios of linear functions subject to a convex feasible set.

Contrary to fractional programming it does not appear to exist a variable transformation which can transform a convex generalized fractional program (P) into a convex program. Consequently, this also complicates the derivation of duality results for convex generalized fractional programs. Observe that for convex fractional programs, duality results can easily be obtained by means of the transformed problem, see Section 3.1. Although more elaborate, a possible way to derive duality results is given by the parametric type of approach. Among the theoretical results, references [6, 40, 42, 79, 133] contain the most important duality theorems in convex generalized fractional programming programming. However, these duality results are basically theoretical and unfortunately no effort was made to use this knowledge for algorithmic purposes. Recently, Barros *et al.* [14, 15] show how duality concepts can actually be used in order to construct efficient computational tools to solve this class of problems. These new algorithms as well as the standard duality results will be discussed in detail in Section 4.2. It is important to mention that lately another dual type of approach has been proposed. Gugat [66] introduce for a special class of convex generalized fractional problems a procedure based on the Fenchel dual of a convex auxiliary problem. This algorithm is an interval-type algorithm and produces a sequence of superlinear convergent points to the optimal solution value.

In Section 4.1 we will survey the Dinkelbach-type approach of Crouzeix *et al.* [43, 44]. Moreover, in Section 4.1.2 we will discuss an allocation problem occurring in continuous location. Unfortunately, this problem corresponds to a nonstandard generalized fractional programming problem, i.e. a generalized fractional programming problem for which the positivity assumption on the denominators has to be imposed in the feasible set. For this class of problems we will present in Section 4.1.3 some special solution techniques based on the primal approach.

4.1   A Primal Approach

The majority of the algorithmic procedures available in the literature only aims to solve the primal problem (P). Among them the most popular is the parametric approach which extends to the multi-ratio case the ideas developed for the single-ratio. In this section, we will start by reviewing the basic results on the parametric procedures of Crouzeix *et al.* [43, 44]. Variants and extensions of this class of algorithms can be found in the excellent survey of Crouzeix and Ferland [41].

One of the standard assumptions of generalized fractional programming is the positivity of the denominators of the ratios of the objective function. However, there exist some problems in real life, like the allocation model presented in Section 4.1.2, for which this assumption has to be included in the feasible set. The inclusion of such constraints not only increases the number of constraints of the feasible set but, may in some cases, destroy "nice" characteristics of the initial feasible set. Therefore, Section 4.1.3 is devoted to developing special solution techniques to tackle this nonstandard class of generalized fractional programs.

4.1.1 The Parametric Approach

The most popular procedure to solve (P) corresponds to solving a sequence of parametric problems given by

$$F(\lambda) = \inf_{x \in \mathcal{X}} \left\{ \max_{j \in J} \{ f_j(x) - \lambda g_j(x) \} \right\}. \qquad (P_\lambda)$$

Similar to fractional programming, problems (P) and (P_λ) are related by the following result.

Lemma 4.1.1 ([43])
Denoting $\vartheta(P)$ by $\lambda_\star$ we have that

(a) $F(\lambda) < +\infty$ *for all* $\lambda \in \mathbb{R}$. *Moreover, the parametric function F is upper semicontinuous and decreasing.*

(b) $F(\lambda) < 0$ *if and only if* $\lambda > \lambda_\star$. *Moreover,* $F(\lambda_\star) \geq 0$.

(c) *If (P) is solvable then* $F(\lambda_\star) = 0$.

(d) *If $F(\lambda_\star) = 0$, then (P) and $(P_{\lambda_\star})$ have the same set of optimal solutions (which may be empty).*

Whenever $\mathcal{X}$ is a compact set the statements in Lemma 4.1.1 can be strengthen. Observe first, that in this case the function ψ is finite-valued and continuous on $\mathcal{X}$ and therefore there exists an optimal solution of (P) in $\mathcal{X}$, i.e. problem (P) is solvable.

Lemma 4.1.2 ([43])

If $\mathcal{X}$ is compact and $n \geq 1$, then

(a) $F(\lambda) < +\infty$, F is strictly decreasing and continuous.

(b) Both (P) and (P_λ) are solvable.

(c) The optimal objective value $\lambda_\star$ of (P) is finite and $F(\lambda_\star) = 0$.

(d) $F(\lambda) = 0$ implies $\lambda = \lambda_\star$.

By Lemma 4.1.2 it is clear that solving (P) is equivalent to finding the unique solution of the univariate equation $F(\lambda) = 0$. Hence, similar to the one ratio case, we are interested in constructing a decreasing sequence λ_k, $k \geq 2$, converging from above to the root of this equation The main idea behind the construction of this sequence

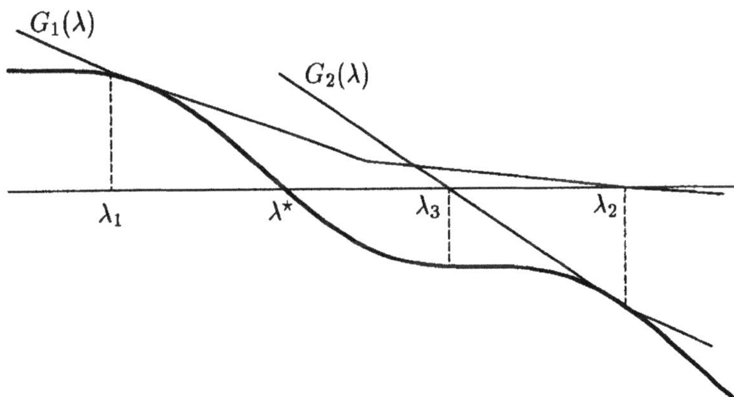

Figure 4.1: Geometric interpretation of the Dinkelbach-type algorithm.

corresponds to approximate the function F iteratively from above by a simpler decreasing function G_k, $k \geq 1$ and to compute for each k the root of the equation $G_k(\lambda) = 0$, see Figure 4.1. Therefore, consider the following approximation of the function F at λ_k, given by $G_k : \mathbb{R} \longrightarrow \mathbb{R}$ with $G_k(\lambda) := \max_{j \in J}\{f_j(x_k) - \lambda g_j(x_k)\}$ and x_k an optimal solution of (P_{λ_k}). Clearly, $G_k(\lambda_k) = F(\lambda_k)$, $G_k(\lambda) \geq F(\lambda)$ and the root of the equation $G_k(\lambda) = 0$ given by $\psi(x_k) = \max_{j \in J} \frac{f_j(x_k)}{g_j(x_k)}$ is an upper bound on λ_*, see Figure 4.1. Moreover, it can easily be shown for $\lambda_k \geq \vartheta(P)$ that $\lambda_{k+1} := \psi(x_k) \leq \lambda_k$.

The Dinkelbach-type algorithm proposed by Crouzeix *et al.* [43] is summarized in Algorithm 4.1.

Step 0. Take $x_0 \in \mathcal{X}$, compute $\lambda_1 := \max_{j \in J} \dfrac{f_j(x_0)}{g_j(x_0)}$ and let $k := 1$;

Step 1. Determine $x_k := \arg \min_{x \in \mathcal{X}} \left\{ \max_{j \in J} \{ f_j(x) - \lambda_k g_j(x) \} \right\}$;

Step 2. If $F(\lambda_k) = 0$

   Then x_k is an optimal solution with value λ_k and **Stop**.

   Else GoTo *Step* 3;

Step 3. Let $\lambda_{k+1} := \max_{j \in J} \dfrac{f_j(x_k)}{g_j(x_k)}$, let $k := k + 1$ and **GoTo** *Step* 1.

Algorithm 4.1: The Dinkelbach-type algorithm.

Notice that for $n = 1$ the above algorithm reduces to the Dinkelbach algorithm described in the previous chapter. Obviously, it is only useful to apply the Dinkelbach-type algorithm if every subproblem (P_λ) is easier to solve than the original problem (P). For instance, for convex generalized fractional programming the parametric problem is a convex problem. Another example is given by generalized linear fractional programming where the associated parametric problems have also a "nice" form since they correspond to linear programming problems. Actually, this algorithm was first drawn by Charnes and Cooper [35] for this class of problems.

Unfortunately, one handicap of this straightforward algorithm is its convergence rate. In fact, while for the special case $n = 1$ Algorithm 4.1 has a superlinear convergence rate, for $n > 1$ Crouzeix *et al.* [43] showed that the convergence rate is only linear. This is mainly due to the fact that the parametric function F is no longer concave.

Theorem 4.1.1 ([43])
If $\mathcal{X}$ is compact then the sequence $\{\lambda_k\}_{k \geq 1}$ generated by the Dinkelbach-type algorithm converges linearly to $\lambda_* = \vartheta(P)$, and each convergent subsequence of $\{x_k\}_{k \geq 1}$ converges to an optimal solution of (P).

Crouzeix *et al.* [43] also present (Proposition 4.1) a condition to ensure local concavity of F. Observe that this implies that the Dinkelbach-type algorithm will behave like Dinkelbach's algorithm ($n = 1$) in a neighborhood of the optimal objective value of (P).

The Dinkelbach-type algorithm can also be applied if $\mathcal{X}$ is not compact. However, in this situation some additional conditions have to be imposed in order to apply this algorithm and to guarantee its convergence. Clearly, Algorithm 4.1 can only be applied if we can guarantee that (P) has a finite solution value and that for $\lambda \geq \vartheta(P)$ the associated parametric problems are solvable. Denoting $\max_{j \in J} g_j(\boldsymbol{x})$ by $\overline{g}(\boldsymbol{x})$ we have the following result.

Corollary 4.1.1 ([43])
If for $\lambda \geq \vartheta(P)$ the associated parametric problems are solvable, (P) is solvable and $\sup_k \overline{g}(\boldsymbol{x}_k) < \infty$ then the sequence $\{\lambda_k\}_{k \geq 1}$ generated by the Dinkelbach-type algorithm converges to $\vartheta(P)$. Moreover, the convergence rate is linear.

The main difference between the Dinkelbach-type algorithm and the Dinkelbach algorithm lies in the way the approximation functions of F are computed. In fact, for the single-ratio case the Dinkelbach algorithm uses a subgradient of the parametric function F to compute an affine upper approximation of this function, see Figure 3.3. To present a related result for the multi-ratio case it is important to introduce the concept of upper subdifferentiability.

Definition 4.1.1 ([106])
A real valued function f defined on $\mathcal{X} \subset \mathbb{R}^m$ is called upper subdifferentiable at $\boldsymbol{x} \in \mathcal{X}$ if there exists a vector $\boldsymbol{u} \in \mathbb{R}^m$ such that

$$f(\boldsymbol{y}) \leq f(\boldsymbol{x}) + (\boldsymbol{y} - \boldsymbol{x})^\top \boldsymbol{u}, \text{ for all } \boldsymbol{y} \in \mathcal{X} \text{ such that } f(\boldsymbol{y}) > f(\boldsymbol{x}).$$

In this case $\boldsymbol{u}$ is called an upper subgradient of f at $\boldsymbol{x}$.

From Lemma 4.1.1 it follows by the monotonicity of the function F that this function is quasiconcave (and quasiconvex). Thus, similar to Lemma 3.1.2, we can in some cases exhibit an upper subgradient of the function F.

Lemma 4.1.3

Let λ_k be such that $F(\lambda_k)$ is finite and the set of optimal solution of (P_{λ_k}) is not empty. Then F is upper subdifferentiable at λ_k and an upper subgradient is given by $-\bar{g}(x_k)$, with x_k an optimal solution of (P_{λ_k}).

Proof: The proof follows easily from Definition 4.1.1 and the definition of F. $\quad\square$

The above lemma suggests the construction of an algorithm using these upper subgradients to derive affine upper approximations of the parametric function F, see Figure 4.2.

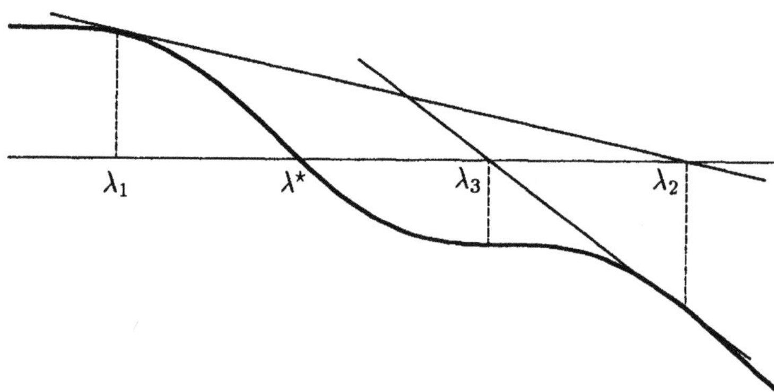

Figure 4.2: An upper subgradient type algorithm.

It is now interesting to analyze the relation between such a procedure and the Dinkelbach-type algorithm.

Lemma 4.1.4

Let λ_k be such that $F(\lambda_k)$ is finite and the set of optimal solution of (P_{λ_k}) is not empty. Then for $\lambda_k \geq \vartheta(P) = \lambda_\star$ we have that

$$\lambda_\star \leq \lambda_{k+1} \leq \frac{F(\lambda_k)}{\bar{g}(x_k)} + \lambda_k \leq \lambda_k$$

with x_k an optimal solution of (P_{λ_k}) and $\lambda_{k+1} := \max_{j \in J} \frac{f_j(x_k)}{g_j(x_k)}$, i.e. the next iteration step of the Dinkelbach-type algorithm.

Proof: Notice first by Lemma 4.1.3 that the inequality

$$F(\lambda_\star) \leq F(\lambda_k) + (\lambda_k - \lambda_\star)\bar{g}(x_k)$$

holds for $\lambda_k \geq \lambda_\star$. Applying now Lemma 4.1.1 and $g_j(x_k) > 0$ for all $j \in J$ it follows from this inequality that

$$\lambda_\star \leq \frac{F(\lambda_k)}{\bar{g}(x_k)} + \lambda_k.$$

Again by Lemma 4.1.1 we obtain, due to $\lambda_k \geq \lambda_\star$, that $F(\lambda_k) \leq 0$ and hence

$$\frac{F(\lambda_k)}{\bar{g}(x_k)} + \lambda_k \leq \lambda_k$$

is verified. Since

$$\frac{F(\lambda_k)}{\bar{g}(x_k)} + \lambda_k = \frac{1}{\bar{g}(x_k)} \max_{j \in J} \left\{ f_j(x_k) - \lambda_k \left(g_j(x_k) - \bar{g}(x_k) \right) \right\}$$

and by Corollary 4.1.1 we have $\lambda_{k+1} \leq \lambda_k$, it follows from the previous inequality that

$$\frac{F(\lambda_k)}{\bar{g}(x_k)} + \lambda_k \geq \frac{1}{\bar{g}(x_k)} \max_{j \in J} \left\{ f_j(x_k) - \lambda_{k+1} g_j(x_k) \right\} + \lambda_{k+1}.$$

By the definition of λ_{k+1} the expression $\max_{j \in J} \left\{ f_j(x_k) - \lambda_{k+1} g_j(x_k) \right\}$ equals zero and so

$$\frac{F(\lambda_k)}{\bar{g}(x_k)} + \lambda_k \geq \lambda_{k+1}.$$

$\square$

Unfortunately, the above lemma shows that this type of approach using an upper subgradient is not worthwhile to pursue, since this procedure will never dominate the Dinkelbach-type algorithm.

Similar to the Dinkelbach algorithm for fractional programming, the Dinkelbach-type algorithm for generalized fractional programming seems to be an ad-hoc procedure. In fact, the results of Lemma 4.1.2 are derived using only algebraic manipulations. However, although not immediately clear, the Dinkelbach-type algorithm can be derived from classical techniques in nonlinear programming. Barros and

Frenk [12] show, using the upper subdifferentiability of the ratio function, that this algorithm corresponds to a special case of a cutting plane algorithm. This new interpretation of the Dinkelbach-type algorithm also enables the identification of lower bounds associated with this well-known "upper bounding" procedure. In particular, the Dinkelbach-type algorithm applied to (P) with $\mathcal{X}$ compact generates not only the decreasing sequence of iteration points λ_k, but also an increasing sequence of lower bounds of the form

$$\frac{1}{\delta}F(\lambda_k) + \lambda_k, \tag{4.2}$$

with $\delta := \min_{x \in \mathcal{X}} \min_{j \in J} g_j(x)$ and this sequence converges from below to $\vartheta(P)$, see [12].

Similar to fractional programming, see Section 3.1.1, expression (4.2) can be used to establish a stopping rule for the Dinkelbach-type algorithm. Accordingly, we can replace $F(\lambda_k) = 0$ in *Step* 2 by

$$|F(\lambda_k)| \leq \varepsilon \delta. \tag{4.3}$$

Clearly, if the algorithm stops at iteration k, using this rule we have that the current approximation λ_k of the optimal value $\vartheta(P)$ is such that $\vartheta(P) \leq \lambda_k \leq \vartheta(P) + \varepsilon$. Observe, due to the monotonicity of the sequence $\{\lambda_k\}_{k \geq 1}$, that the above relation is also verified for $\lambda_{k+1} := \psi(x_k)$ with x_k an optimal solution of (P_{λ_k}) and so $\vartheta(P) \leq \lambda_{k+1} \leq \vartheta(P) + \varepsilon$.

The above result and Lemma 4.1.3 are related to the next proposition. In particular, (4.5) of Proposition 4.1.1 implies (4.2), while Lemma 4.1.3 provides a clear geometrical interpretation of (4.4).

Proposition 4.1.1 ([43])
Let λ_k be such that $F(\lambda_k)$ is finite and the set of optimal solution of (P_{λ_k}) is not empty. Then,

$$F(\lambda) \leq F(\lambda_k) + (\lambda_k - \lambda)\overline{g}(x_k) \qquad if\ \lambda < \lambda_k, \tag{4.4}$$

$$F(\lambda) \leq F(\lambda_k) + (\lambda_k - \lambda)\underline{g}(x_k) \qquad if\ \lambda > \lambda_k, \tag{4.5}$$

with x_k an optimal solution of (P_{λ_k}) and $\underline{g}(x) := \min_{j \in J} g_j(x)$.

A refinement of the Dinkelbach-type algorithm was later proposed by Crouzeix *et al.* [44] and independently by Flachs [52]. The main idea behind this modified procedure consists in trying to "smooth" the parametric function. This is achieved by scaling this parametric function in such a way that near the optimum value the function F will be concave. In order to achieve this Crouzeix *et al.* [44] propose the following reformulation of (P)

$$\inf_{x \in \mathcal{X}} \max_{j \in J} \frac{f_j(x)/g_j(x_\star)}{g_j(x)/g_j(x_\star)} \qquad (P')$$

where $x_\star$ denotes an optimal solution of (P). The associated parametric problem is given by

$$F'(\lambda) = \inf_{x \in \mathcal{X}} \left\{ \max_{j \in J} \left\{ \frac{f_j(x) - \lambda g_j(x)}{g_j(x_\star)} \right\} \right\}. \qquad (P'_\lambda)$$

Applying now Proposition 4.1.1 yields

$$F'(\lambda) \leq F'(\lambda_\star) + (\lambda_\star - \lambda)\overline{\rho}(x_\star) \qquad \text{if } \lambda < \lambda_\star$$

$$F'(\lambda) \leq F'(\lambda_\star) + (\lambda_\star - \lambda)\underline{\rho}(x_\star) \qquad \text{if } \lambda > \lambda_\star$$

with

$$\underline{\rho}(x_\star) := \min_{j \in J} \frac{g_j(x_\star)}{g_j(x_\star)} = 1 = \max_{j \in J} \frac{g_j(x_\star)}{g_j(x_\star)} =: \overline{\rho}(x_\star).$$

The above inequalities reduce to a single one, implying that the function F' is concave on the neighborhood of $\lambda_\star$. Hence, in a neighborhood of the optimum, the Dinkelbach-type algorithm "coincides" with Newton's algorithm, and thus its convergence rate is superlinear.

In practice, since an optimal solution of (P) is not known a priori, the current iteration point x_{k-1} is used instead. Observe that this point corresponds to the best approximation of $x_\star$ available. Hence, the Dinkelbach-type-2 corresponds to Algorithm 4.1 where *Step* 1 is replaced by

Step $1'$. Determine:

$$x_k := \arg\min_{x \in \mathcal{X}} \left\{ \max_{j \in J} \left\{ \frac{f_j(x) - \lambda_k g_j(x)}{g_j(x_{k-1})} \right\} \right\}$$

Borde and Crouzeix [29] discuss the convergence rate of this so called Dinkelbach-type-2 algorithm and show that under stronger assumptions it is possible to achieve superlinear convergence, or even to obtain a quadratic convergence rate.

Based on the Dinkelbach-type approaches and their geometrical interpretation, several interval type algorithms have been proposed. Ferland and Potvin [49] discuss these interval type algorithms and provide computational experience for the linear case. Also convergence rate results for these type of algorithms are presented by Bernard and Ferland [23]. A thorough overview of these algorithms can be found in the survey by Crouzeix and Ferland [41].

The above Dinkelbach-type approaches can be applied to all types of generalized fractional programs, including integer generalized fractional programs. However, in the nonconvex case the parametric problems to be solved become too "complicated", and this probably justifies why most of the research is devoted to the convex case.

So far, we have only analyzed the case where the generalized fractional programming problems satisfy the usual positivity assumption, i.e. the denominators of the ratios of the objective function are positive on the feasible set $\mathcal{X}$. In the next section we will consider an example of a problem not satisfying this standard assumption and discuss how the Dinkelbach-type approach described in this section can be efficiently "adapted" for this type of problems.

4.1.2   An Allocation Model

In the previous chapters we have focused on some discrete location models. For these models there exists a simple rule to assign the clients to the facilities. In particular, for the models discussed in Chapter 2 the clients are assigned to the most profitable facility. However, if clients from a given demand point are generated according to a stochastic process evolving in time, it may be impossible to assign all these clients to the same facility due to work load considerations. This means that these clients may have to use the services of different facilities. Observe that the work load of each facility can be incorporated into the objective function and to model this we may use queueing theory. One of the first references using queueing theory in location models is given by Berman *et al.* [22]. Later on, several variants of this model have been considered like the model due to Batta [19] where clients can be rejected.

In this section we will discuss an allocation problem associated with the *multi-facility location queueing problem* where the objective function involves the work load of the facilities. It is therefore assumed that the facilities are already located, and hence it is necessary to determine the best policy of assigning "incoming" clients to these existing facilities. To keep the model mathematically tractable we assume that at each facility there is one server who travels at unit speed to a customer and after servicing him/her, this server returns to the facility.

An example of such a model is given by an urban system consisting of ambulance depots. In this case the server is the ambulance bringing the customer to a hospital near this depot. If a demand from a client comes in by phone, this client will be directed to one of the facilities in such a way that the work loads among the existing facilities become balanced. Without loss of generality the on-scene and off-scene service times are ignored and so the total service time only depends on the distance between the clients and the facilities In this model, a logical criterion is to allocate the clients among the existing facilities, so as to minimize the maximum of the expected work loads (in the steady state).

In order to introduce this allocation model, let $I := \{a_1, \ldots, a_m\} \subseteq I\!\!R^2$ denote the set of m different demand points and $J := \{b_1, \ldots, b_n\} \subseteq I\!\!R^2$ the set of n different locations of identical facilities. Moreover, the sum of the distances from facility j at b_j, $j = 1, \ldots, n$, to the demand point i at a_i, $i = 1, \ldots, m$, and from demand point i to facility j is denoted by the positive constant d_{ij}. It is assumed that each demand point i generates calls according to a Poisson process $\underline{P}_i(t), t \geq 0$ with rate $\lambda_i > 0$ and that the Poisson processes $\underline{P}_1(t), \ldots, \underline{P}_m(t)$ are independent. In the remainder, calls generated by demand point i are called type-i calls. If a call from one of the demand points is assigned to facility j, it waits for service from that facility. It is assumed that each facility has only one server and that the queueing discipline is work-conserving and nonpreemptive. This means that the server is not idle if the queue of assigned calls is nonempty and that the queueing discipline does not affect either the amount of service time given to a call or the arrival time of any call. Moreover, it also implies that once a service is started to a given call, this service will be completed. Well-known examples of work-conserving and nonpreemptive queueing disciplines are FCFS (First-Come-First-Served) and random order of service. Once the server starts to serve a given call this server travels at unit speed from facility j to the demand point which generated this call and returns after servicing to the home facility j. Contrary to Berman *et al.* [22]

we will assume, that the on-scene and off-scene service times can be ignored and so the total service time s_{ij} given to a type-i call by the server of facility j equals d_{ij}. In order to determine the best random assignment policy of calls to facilities it is necessary to introduce random policies $x \in I\!\!R^{mn}$ consisting of components x_{ij}, for $i \in I$ and $j \in J$ with

$$x_{ij} := \Pr\{\text{type-}i \text{ call is assigned to facility } j\}.$$

Clearly, $0 \leq x_{ij} \leq 1$ and $\sum_{j=1}^{n} x_{ij} = 1$ for every $i \in I$. Since the objective is to select a random policy that minimizes the maximum of the average work load at each facility $j \in J$, we denote by $W_j(x)$ the average amount of unfinished work in the steady state (due to the assigned calls) for the server at facility j if the random policy x is used. In [12] it was shown that this average amount of unfinished work $W_j(x)$, also called the average work-in-system, is given by:

$$W_j(x) = \begin{cases} \dfrac{\frac{1}{2}\sum_{i=1}^{m} \lambda_i s_{ij}^2 x_{ij}}{1 - \sum_{k=1}^{m} \lambda_k s_{kj} x_{kj}} & \text{if } 1 - \sum_{k=1}^{m} \lambda_k s_{kj} x_{kj} > 0 \\ +\infty & \text{otherwise} \end{cases}.$$

Introducing the compact set $\mathcal{X} \subset I\!\!R^{mn}$ of random policies , i.e. $\mathcal{X} := \{x \in I\!\!R^{mn} : x_{ij} \geq 0, \sum_{j=1}^{n} x_{ij} = 1$ for every $i \in I\}$ this allocation problem boils down to the generalized linear fractional programming problem

$$\min_{x \in \mathcal{X}_c} \max_{j \in J} W_j(x) \qquad\qquad (Alloc)$$

with

$$\mathcal{X}_c := \left\{ x \in \mathcal{X} : 1 - \sum_{i=1}^{m} \lambda_i s_{ij} x_{ij} > 0, \text{ for every } j \in J \right\}.$$

Observe, if the system is totally stable, i.e. each facility can handle all the clients, that the feasible set of ($Alloc$) can be replaced by $\mathcal{X}$. In this case, ($Alloc$) corresponds to a standard generalized fractional problem and therefore any of the solution techniques discussed in the previous section can be applied. However, the totally stability assumption is too restrictive and hence it is important to analyze the general case. The introduction of the constraints ensuring the positivity of the denominators in the feasible set creates a "problem" in the sense that it spoils the structure of the original feasible $\mathcal{X}$. Moreover, it also increases the number of constraints in this set. Hence, it is important to construct special solution techniques that can handle in a efficient way the inclusion of the positivity assumptions in the feasible set. Actually, the optimization problem (*Alloc*) corresponds to a nonstandard generalized fractional programming problem and hence the next section is devoted to the analysis of such nonstandard generalized fractional programming problems.

In the above discussion of the model it is assumed that the location of the facilities is known. Although this assumption is quite reasonable in some cases, it is also interesting to briefly analyze the associated location-allocation problem. This type of problems is characterized by the fact that both the location of facilities and the assignment of the clients have to be simultaneously decided, and thus is a difficult problem. One of the first references discussing solution procedures for location-allocation problems is given by Cooper [36]. Cooper [37] proposed one of the most applied heuristics for this type of problems which consists simply in decomposing the problem into the location problem and the allocation problem. Due to its simplicity this decomposition scheme is very popular. Observe that in our case the problem of locating n facilities in the plane for some allocation policy also belongs to the fractional programming class. In fact, denoting by $y \in \mathbb{R}^{2n}$ the vector containing the coordinates of all possible locations $y_j \in \mathbb{R}^2$, and the assigned policy of calls to facilities by $p \in \mathbb{R}^{mn}$, this problem corresponds to

$$\inf_{y \in \mathcal{Y}} \left\{ \max_{j \in J} \frac{\frac{1}{2} \sum_{i \in I} \lambda_i p_{ij} s_i(y_j)^2}{1 - \sum_{i \in I} \lambda_i p_{ij} s_i(y_j)} \right\} \tag{Loc}$$

with

$$\mathcal{Y} := \left\{ y \in \mathbb{R}^{2n} : 1 - \sum_{i \in I} \lambda_i p_{ij} s_i(y_j) > 0, \text{ for every } j \in J \right\}$$

and $s_i(y_j)$ a function expressing the service time for demand point i if facility j is located at y_j. At first sight the above optimization problem appears to be a generalized fractional programming problem. However, due to the special "separable" form of the feasible set, the above problem can be related to the following fractional programming problem given by

$$\inf_{y_j \in \mathcal{Y}_j} \left\{ \frac{\frac{1}{2} \sum_{i \in I} \lambda_i p_{ij} s_i(y_j)^2}{1 - \sum_{i \in I} \lambda_i p_{ij} s_i(y_j)} \right\} \tag{Loc_j}$$

with

$$\mathcal{Y}_j := \left\{ y_j \in \mathbb{R}^2, 1 - \sum_{i \in I} \lambda_i p_{ij} s_i(y_j) > 0 \right\}.$$

Clearly, $\vartheta(Loc) \geq \max_{j \in J} \vartheta(Loc_j)$. Due to $\mathcal{Y} = \mathcal{Y}_1 \times \ldots \times \mathcal{Y}_n$, it follows that $\vartheta(Loc) \leq \max_{j \in J} \vartheta(Loc_j)$. In fact, if $\max_{j \in J} \vartheta(Loc_j) < \vartheta(Loc)$ then an optimal solution of $\max_{j \in J} \vartheta(Loc_j)$ will also be feasible to (Loc) with a smaller objective value than $\vartheta(Loc)$ which yields a contradiction. Hence, solving (Loc) is equivalent to solving the n fractional programming problems (Loc_j). Observe that, due to the form of the feasible set, the optimization problems (Loc_j) can also be seen as nonstandard fractional programming problems. Hence, the approach described in the next section to handle the inclusion of the positivity assumptions in the feasible set of a generalized fractional program can also easily be adapted for the single-ratio case.

4.1.3 A Nonstandard Class of Generalized Fractional Programs

As a working hypothesis, it is always assumed that the denominators of the ratios of the objective function are positive on the considered domain. However, for some practical applications, like the allocation model discussed in the previous section, the feasible set may include points at which the denominators of the ratios are not positive. In this section we will consider a class of generalized fractional programs for which the positivity assumption is part of the feasible set. To simplify the notation, we will discuss this problem in the general framework of this chapter. Hence, we will assume that the feasible set $\mathcal{X} \subset \mathbb{R}^m$ is compact and $f_j, g_j : \mathcal{C} \longrightarrow \mathbb{R}$ are continuous functions for all $j \in J := \{1, \ldots, n\}$ on the open set $\mathcal{C}$ containing $\mathcal{X}$. Moreover, let $\mathcal{X}_c := \{x \in \mathcal{X} : g_j(x) > 0 \text{ for every } j \in J\}$. We will now consider the generalized fractional program given by

$$\inf \{\psi(x) : x \in \mathcal{X}_c\} \qquad (P_c)$$

with the function ψ defined in (4.1). We will also need to impose the following additional assumptions

(a) $f_j(x) \geq 0$ for every $x \in \mathcal{X}$ and $j \in J$.

(b) If $f_j(x) = 0$ for some $x \in \mathcal{X}$ and $j \in J$ then $g_j(x) > 0$.

Clearly, the allocation problem (*Alloc*) belongs to the above class of optimization problems.

For the optimization problem (P_c), the feasible set $\mathcal{X}_c$ may be empty. By the compactness of $\mathcal{X}$ and the continuity of g_j, $j \in J$, $\mathcal{X}_c$ is nonempty if and only if $\max_{x \in \mathcal{X}} \min_{j \in J} g_j(x) > 0$. Hence, we will assume that some $\widehat{x} \in \mathcal{X}_c$ is known. By Assumption (a) it follows that $\psi(x) \geq 0$ for every $x \in \mathcal{X}_c$ and hence we may assume without loss of generality that $\widehat{\lambda} := \psi(\widehat{x}) > 0$. Although the set $\mathcal{X}_c$ is in general not closed, it will be shown in Lemma 4.1.5 that (P_c) is solvable. It will also be proved in Lemma 4.1.6 that the associated parametric problem

$$F_c(\lambda) := \inf_{x \in \mathcal{X}_c} \left\{ \max_{j \in J} \left\{ f_j(x) - \lambda g_j(x) \right\} \right\} \qquad (P_{c_\lambda})$$

is solvable for $\lambda \geq \vartheta(P_c)$. This implies by Corollary 4.1.1 that the Dinkelbach-type algorithm can be applied to (P_c) and has a linear convergence rate. However, since $\mathcal{X}_c$ is in general not closed it may be difficult to solve each subproblem (P_{c_λ}) by standard methods. For instance, if f_j, g_j, $j \in J$ are affine and $\mathcal{X}$ is defined by linear constraints the subproblem (P_{c_λ}) is no longer a linear programming problem.

Instead of following the above classical approach to solve (P_c) we will show that it is also possible to apply the same parametric approach to a "smaller" subproblem. In this subproblem the constraints on the denominator are dropped from the feasible set. In order to prove the first lemma we introduce the nonempty compact set $\widehat{\mathcal{X}} \subset \mathbb{R}^m$ given by

$$\widehat{\mathcal{X}} := \left\{ x \in \mathcal{X} : f_j(x) - \widehat{\lambda} g_j(x) \leq 0 \text{ for every } j \in J \right\} \qquad (4.6)$$

and consider the optimization problem

$$\inf \left\{ \psi(x) : x \in \widehat{\mathcal{X}} \right\}, \qquad (\widehat{P})$$

and its associated parametric problem

$$\widehat{F}(\lambda) := \inf_{x \in \widehat{\mathcal{X}}} \left\{ \max_{j \in J} \left\{ f_j(x) - \lambda g_j(x) \right\} \right\}. \qquad (\widehat{P}_\lambda)$$

The following result relates problems (P_c) and $(\widehat{P})$.

Lemma 4.1.5

The optimal solution sets of (P_c) and $(\widehat{P})$ coincide and are nonempty. Moreover, $\vartheta(P_c)$ equals $\vartheta(\widehat{P})$.

Proof: To prove the first part it is enough to verify that

$$\widehat{\mathcal{X}} = \left\{ x \in \mathcal{X}_c : \psi(x) \leq \widehat{\lambda} \right\}.$$

Clearly, for some $x \in \mathcal{X}_c$ satisfying $\psi(x) \leq \widehat{\lambda}$ it follows that $x \in \widehat{\mathcal{X}}$. Also, if $x \in \widehat{\mathcal{X}}$ we obtain by Assumption (a) that $f_j(x) \geq 0$. If $f_j(x) = 0$ this yields by Assumption (b) that $g_j(x) > 0$. Moreover, since $\widehat{\lambda} > 0$ we obtain for $f_j(x) > 0$ that $g_j(x) \geq \widehat{\lambda}^{-1} f_j(x) > 0$ and thus $x \in \mathcal{X}_c$. Finally, for $x \in \mathcal{X}_c$ it follows that $f_j(x) - \widehat{\lambda}g_j(x) \leq 0$ for every $j \in J$ if and only if $\psi(x) \leq \widehat{\lambda}$ and this proves the first result.

To prove the second part observe, by the continuity of ψ on $\{x \in \mathcal{C} : g_j(x) > 0, j \in J\}$, the compactness of $\widehat{\mathcal{X}}$ and $\widehat{x} \in \widehat{\mathcal{X}}$, that the optimal solution set of $(\widehat{P})$ is nonempty. This implies by the first result that the optimal solution set of (P_c) is also nonempty. The last result follows now immediately. $\qquad\square$

We still have to show that (P_{c_λ}) is solvable for $\lambda \geq \vartheta(P_c)$. This is achieved by showing that the parametric problem

$$F(\lambda) := \min_{x \in \mathcal{X}} \left\{ \max_{j \in J} \{f_j(x) - \lambda g_j(x)\} \right\} \qquad (P_\lambda)$$

has the same set of optimal solutions as (P_{c_λ}) for $\lambda \geq \vartheta(P_c)$.

Lemma 4.1.6

For every $\lambda \geq \vartheta(P_c)$ it follows that x_λ is an optimal solution of (P_λ) if and only if x_λ is an optimal solution of (P_{c_λ}). Moreover, both sets of optimal solutions are nonempty and $F(\lambda)$ equals $F_c(\lambda)$ for every $\lambda \geq \vartheta(P_c)$.

Proof: From Lemma 4.1.5 we have that

$$\widehat{\mathcal{X}} = \left\{ x \in \mathcal{X}_c : \psi(x) \leq \widehat{\lambda} \right\}.$$

Hence, by applying Lemma 4.1.2 to $(\widehat{P})$ and Lemma 4.1.5 we obtain that $\widehat{F}(\vartheta(P_c)) = 0$ and $\widehat{F}$ is decreasing. Since $\widehat{\mathcal{X}} \subset \mathcal{X}$ this implies that $F(\lambda) \leq \widehat{F}(\lambda) \leq 0$ for every $\lambda \geq \vartheta(P_c)$. If $\boldsymbol{x}_\lambda \in \mathcal{X}$ is an optimal solution of (P_λ) the above inequality yields $f_j(\boldsymbol{x}_\lambda) - \lambda g_j(\boldsymbol{x}_\lambda) \leq 0$ for every $j \in J$ and $\lambda \geq \vartheta(P_c) \geq 0$. Hence, if $f_j(\boldsymbol{x}_\lambda) > 0$ for some $j \in J$ then necessarily $g_j(\boldsymbol{x}_\lambda) > 0$. Also, if $f_j(\boldsymbol{x}_\lambda) = 0$ then by Assumption (b) we obtain that $g_j(\boldsymbol{x}_\lambda) > 0$. This shows that $\boldsymbol{x}_\lambda \in \mathcal{X}_c$ and using $\mathcal{X}_c \subseteq \mathcal{X}$ it must follow that $\boldsymbol{x}_\lambda$ is also an optimal solution of (P_{c_λ}).

To prove the reverse implication we first observe by the continuity of $f_j, g_j, j \in J$ and the compactness of $\mathcal{X}$ that the optimization problem (P_λ) has an optimal solution $\boldsymbol{y}_\lambda$. Hence, by the first part, $\boldsymbol{y}_\lambda \in \mathcal{X}_c$ and so for $\boldsymbol{x}_\lambda$ an optimal solution of (P_{c_λ}) it must follow that

$$\max_{j \in J}\{f_j(\boldsymbol{x}_\lambda) - \lambda g_j(\boldsymbol{x}_\lambda)\} \leq \max_{j \in J}\{f_j(\boldsymbol{y}_\lambda) - \lambda g_j(\boldsymbol{y}_\lambda)\}.$$

This yields that $\boldsymbol{x}_\lambda$ is also an optimal solution of (P_λ) and hence the first part of the lemma is proved. Using this result the last part of this lemma follows immediately.

$\square$

By Lemmas 4.1.5 and 4.1.6 it is obvious that (P_c) can be solved by applying the Dinkelbach-type algorithm as described in Algorithm 4.2.

Step 0. Choose $\boldsymbol{x}_0 := \widehat{\boldsymbol{x}} \in \mathcal{X}_c$, let $\lambda_1 := \widehat{\lambda} = \psi(\boldsymbol{x}_0)$ and let $k := 1$;

Step 1. Determine $\boldsymbol{x}_k := \arg\min_{\boldsymbol{x} \in \mathcal{X}}\left\{\max_{j \in J}\{f_j(\boldsymbol{x}) - \lambda_k g_j(\boldsymbol{x})\}\right\}$;

Step 2. **If** $F(\lambda_k) = 0$

        Then $\boldsymbol{x}_k$ is an optimal solution with value λ_k and **Stop**.

        Else GoTo *Step 3*;

Step 3. Let $\lambda_{k+1} := \max_{j \in J} \dfrac{f_j(\boldsymbol{x}_k)}{g_j(\boldsymbol{x}_k)}$; let $k := k + 1$, and **GoTo** *Step 1*.

Algorithm 4.2: Adapted Dinkelbach-type algorithm.

Although the optimization problem (P_c) with a noncompact feasible set can directly be solved by the Dinkelbach-type algorithm described in Section 4.1.1, the above approach considerably simplifies the feasibility set of the subproblems by deleting the nonnegative constraints on the denominators. This observation clearly improves the applicability of the Dinkelbach-type algorithm to this class of problems.

This primal parametric approach has been successfully tested in the case of the allocation problem (*Alloc*), see [33].

4.2 A Dual Approach

For the special class of convex generalized fractional programming problems, the references on duality results are abundant (see [6, 40, 42, 79, 133]). However, these duality results have not been directly used to derive another type of algorithms for this class of problems. In fact, the algorithms to solve convex generalized fractional programs are "primal" algorithms which do not attempt to solve the associated standard dual problem. This may be justified by the fact that the standard dual of a convex generalized fractional program looks much more difficult to handle than its primal counterpart. An exception is given by generalized linear fractional programs for which the dual under certain conditions is again a generalized linear fractional program, see [42, 43, 79] and Section 4.2.1. This led Crouzeix *et al.* [43] to consider solving the dual problem via the Dinkelbach-type algorithm, whenever the unbounded feasible set $\mathcal{X}$ makes it impractical to solve the primal problem directly.

In this section we start by reviewing the standard dual problem of a convex generalized fractional program and discussing in detail the "dual" algorithm presented by Barros *et al.* [15]. As we shall see in Section 4.2.1, this "dual" algorithm extends to the nonlinear case the Dinkelbach-type algorithm of Crouzeix *et al.* [43] applied to the dual problem of a generalized linear fractional program.

The new duality approach introduced by Barros *et al.* [14] is discussed in detail in Section 4.2.2. Using this approach, a new dual for a convex generalized fractional programming problem is presented as well as an efficient algorithm to solve this new dual.

Finally, computational results performed for quadratic-linear ratios and linear constraints, comparing the performance of these two dual type algorithms with the Dinkelbach-type algorithm are provided in Section 4.2.3.

4.2.1   Solving the Standard Dual

In order to present the standard dual of a convex generalized fractional programming problem (P), we will assume that the continuous functions $f_j, g_j : \mathcal{C} \longrightarrow \mathbb{R}$, $j \in J$ are respectively convex and concave on $\mathcal{S}$ with $\mathcal{S} \subset \mathcal{C}$ a compact convex set. In addition, g_j are positive on $\mathcal{X}$ and either the functions f_j are nonnegative on $\mathcal{X}$ or the functions g_j are affine for every $j \in J$ on $\mathcal{S}$. Moreover, the feasible nonempty set $\mathcal{X}$ is given by $\mathcal{X} := \{x \in \mathcal{S} : h(x) \leq 0\}$, and $h : \mathbb{R}^m \longrightarrow \mathbb{R}^r$ is a vector-valued convex function. Clearly, under these conditions the set $\mathcal{X}$ is compact and convex, and therefore (P) has a finite value and is solvable. In order to simplify the notation, we will introduce $f(x)^\top := (f_1(x), \ldots, f_n(x))$ and $g(x)^\top := (g_1(x), \ldots, g_n(x))$.

An easy direct approach to derive the dual problem of (P) is given by Jagannathan and Schaible [79]. Due to $\mathcal{S}$ compact, one can apply the generalized Farkas lemma of Bohnenblust *et al.* [27] to a system of convex inequalities. This leads to the standard dual problem of (P) given by

$$\sup \left\{ t : t y^\top g(x) < y^\top f(x) + z^\top h(x), t \in \mathbb{R}, x \in \mathcal{S}, y \in \Sigma, z \geq 0 \right\}$$

with $\Sigma := \{ y \in \mathbb{R}^n : y \geq 0, \sum_{j \in J} y_j = 1 \}$. Clearly, if additionally $g_j(x) > 0$ for every $x \in \mathcal{S}$ and $j \in J$ the above problem can be rewritten as

$$\sup_{y \in \Sigma, z \geq 0} \left\{ \inf_{x \in \mathcal{S}} \frac{y^\top f(x) + z^\top h(x)}{y^\top g(x)} \right\}. \tag{D}$$

Moreover, since the set $\mathcal{S}$ is compact and $y^\top g(x) > 0$ for every $x \in \mathcal{S}$, one may replace inf by min. In the remaining of this section we will always assume that $g_j(x) > 0$ for every $x \in \mathcal{S}$ and $j \in J$. Although $\vartheta(D)$ equals $\vartheta(P)$, see [42, 79], there may not exist an optimal dual solution. Since we are interested in an algorithm to solve this dual problem, we need to guarantee that (D) is solvable. Hence, we will impose a Slater-type condition on the set $\mathcal{X}$.

Slater's condition

Let R denote the set of indices $1 \leq l \leq r$ such that the lth component h_l of the vector-valued convex function $h : \mathbb{R}^m \to \mathbb{R}^r$ is affine and suppose there exists some x belonging to the relative interior ri$(\mathcal{S})$ *of $\mathcal{S}$ satisfying $h_l(x) < 0, l \notin R$ and $h_l(x) \leq 0, l \in R$.*

Using now the indirect Lagrangian approach of Craven [40], the following result is easy to show. Recall that by our assumptions that $\vartheta(P)$ is finite and $\vartheta(P)$ is nonnegative if g is a concave vector-valued function.

Proposition 4.2.1

If the Slater condition holds then the parametric problem

$$\sup_{y \in \Sigma, z \geq 0} \left\{ \min_{x \in \mathcal{S}} \left\{ y^{\mathsf{T}} f(x) + z^{\mathsf{T}} h(x) - \lambda y^{\mathsf{T}} g(x) \right\} \right\} \tag{D_λ}$$

is solvable and $\vartheta(D_\lambda) = \vartheta(P_\lambda)$ for any λ if g is an affine vector-valued function or for $\lambda \geq 0$ if g is a concave vector-valued function. Moreover, the dual problem of (P) is solvable and $\vartheta(D) = \vartheta(P)$.

Proof: By Theorem 28.2 of [117] the Lagrangian dual (D_λ) of the parametric problem (P_λ) given by

$$\sup_{y \in \Sigma, z \geq 0} \left\{ \min_{x \in \mathcal{S}} \left\{ y^{\mathsf{T}} f(x) + z^{\mathsf{T}} h(x) - \lambda y^{\mathsf{T}} g(x) \right\} \right\}$$

is solvable for every $\lambda \in \mathbb{R}$ if the functions g_j are affine or for every $\lambda \geq 0$ if the functions g_j are concave, $j \in J$. It also follows that $\vartheta(D_\lambda) = \vartheta(P_\lambda)$ and this proves the first result.

Using the above result and Lemma 4.1.2 we have for $\lambda_\star = \vartheta(P)$ that

$$0 = \vartheta(P_{\lambda_\star}) = \vartheta(D_{\lambda_\star}) = \max_{y \in \Sigma, z \geq 0} \left\{ \min_{x \in \mathcal{S}} \left\{ y^{\mathsf{T}} f(x) + z^{\mathsf{T}} h(x) - \lambda_\star y^{\mathsf{T}} g(x) \right\} \right\}. \tag{4.7}$$

Hence, there exists some $y_\star \in \Sigma$ and $z_\star \geq 0$ such that

$$\min_{x \in \mathcal{S}} \left\{ y_\star^{\mathsf{T}} f(x) + z_\star^{\mathsf{T}} h(x) - \lambda_\star y_\star^{\mathsf{T}} g(x) \right\} = 0$$

and

$$\min_{x \in \mathcal{S}} \left\{ y^{\mathsf{T}} f(x) + z^{\mathsf{T}} h(x) - \lambda_\star y^{\mathsf{T}} g(x) \right\} \leq 0 \tag{4.8}$$

for every $y \in \Sigma$ and $z \geq 0$. By the above equality and $g(x) > 0$ for every $x \in \mathcal{S}$ we obtain that

$$\lambda_\star = \min_{x \in \mathcal{S}} \left\{ \frac{y_\star^{\mathsf{T}} f(x) + z_\star^{\mathsf{T}} h(x)}{y_\star^{\mathsf{T}} g(x)} \right\}.$$

On the other hand, from (4.8) it follows that

$$\lambda_\star \geq \min_{x \in S} \left\{ \frac{y^\top f(x) + z^\top h(x)}{y^\top g(x)} \right\}$$

for every $y \in \Sigma$ and $z \geq 0$. Hence, by the previous equality and inequality the dual problem (D) is solvable and $\lambda_\star = \vartheta(P)$ equals $\vartheta(D)$. $\qquad\qquad\qquad\square$

We will assume from now on that the Slater condition holds and hence we can rewrite the standard dual (D) as

$$\max_{y \in \Sigma, z \geq 0} d(y, z)$$

where the function $d : \Sigma \times \mathbb{R}_+^r \longrightarrow \mathbb{R}$ is given by

$$d(y, z) := \min_{x \in S} \frac{y^\top f(x) + z^\top h(x)}{y^\top g(x)} \qquad\qquad (4.9)$$

with $\mathbb{R}_+^r$ denoting the nonnegative orthant of $\mathbb{R}^r$. Notice that (4.9) corresponds to a single-ratio fractional programming problem. Furthermore, if the functions g_j are affine for every $j \in J$ then (4.9) corresponds to a convex fractional programming problem. Clearly, by the positivity of g on S and S compact the function d is continuous on $\Sigma \times \mathbb{R}_+^r$, see Lemma 3.1 of [108]. Moreover, the function d is semistrictly quasiconcave since it is the infimum of semistricly quasiconcave functions

$$x \longmapsto \frac{y^\top f(x) + z^\top h(x)}{y^\top g(x)},$$

see [6]. Hence, (D) corresponds to a quasiconcave optimization problem, where a local maximum is a global maximum, see [6]. Notice, since the Slater condition holds, we know that this maximum is attained, i.e. there exists some $(y, z) \in \Sigma \times \mathbb{R}_+^r$ such that $d(y, z) = \vartheta(D)$.

Proposition 4.2.1 also suggests that solving a sequence of subproblems (D_λ) might recover $\vartheta(P)$ and possibly a primal solution. Hence, we will introduce the value function $G : \mathbb{R} \longrightarrow \mathbb{R}$ associated with (D_λ) given by

$$G(\lambda) := \max_{y \in \Sigma, z \geq 0} G(y, z, \lambda)$$

with

$$G(y, z, \lambda) := \min_{x \in S} \left\{ y^\top f(x) + z^\top h(x) - \lambda y^\top g(x) \right\}.$$

It is not difficult to verify that for any $(y, z) \in \Sigma \times I\!\!R_+^r$ the function $G_{(y_k, z_k)}$: $I\!\!R \longrightarrow I\!\!R$ given by $G_{(y_k, z_k)}(\lambda) = G(y_k, z_k, \lambda)$ satisfies

$$G_{(y_k, z_k)}(\lambda) \leq F(\lambda).$$

Moreover, by Proposition 4.2.1 it follows that the function G equals the function F and so if (y_k, z_k) solves (D_{λ_k}) we obtain that

$$G_{(y_k, z_k)}(\lambda_k) = G(\lambda_k) = F(\lambda_k).$$

Hence, the function $\lambda \longmapsto G_{(y_k, z_k)}(\lambda)$ is a concave lower approximation of the function F and this observation suggests that finding the zero of the equation $F(\lambda) = 0$ can be achieved by computing the root of the equation $G_{(y_k, z_k)}(\lambda) = 0$, which is given by $d(y_k, z_k)$. As mentioned before, computing $d(y_k, z_k)$ corresponds to solving a single-ratio fractional programming problem, which can easily be done using the Dinkelbach algorithm described in Section 3.1.1. However, the efficiency of this procedure depends mostly on whether the associated parametric problem has a "nice" form. Clearly, if the g_j are affine functions for every $j \in J$, then the associated parametric problem corresponds to a convex problem. On the other hand, if the functions g_j are concave then the parametric problem is convex only if the parameter λ is nonnegative. Nevertheless, in this case $\vartheta(P) \geq 0$ and hence we are only interested in $(y, z) \in \Sigma \times I\!\!R_+^r$ such that $d(y, z) \geq 0$. Observe that, in this case (4.9) is also a convex single-ratio fractional programming problem. We can now introduce Algorithm 4.3 to solve (D).

In Algorithm 4.3 it is not clear whether the iteration point values $d(y_k, z_k)$, $k \geq 1$ are strictly increasing. Hence, it is important to check for a nonoptimal iteration point (y_{k-1}, z_{k-1}), whether the next iteration point (y_k, z_k) (an optimal solution of $(D_{d(y_{k-1}, z_{k-1})})$) belongs to the upper level set

$$\mathcal{U}_d^\circ(d(y_{k-1}, z_{k-1})) := \left\{ (y, z) \in \Sigma \times I\!\!R_+^r : d(y, z) > d(y_{k-1}, z_{k-1}) \right\}.$$

Introducing also

$$\mathcal{U}_d(d(y_{k-1}, z_{k-1})) := \left\{ (y, z) \in \Sigma \times I\!\!R_+^r : d(y, z) \geq d(y_{k-1}, z_{k-1}) \right\},$$

the following result justifies the choice of (y_k, z_k) used in *Step* 1 of Algorithm 4.3.

Step 0. **If** g_j for all $j \in J$ are concave

        Then Let $\lambda_0 := 0$;

        Else Take $y_0 \in \Sigma, z_0 \geq 0$ and compute

$$\lambda_0 := d(y_0, z_0) = \min_{x \in S} \frac{y_0^{\mathsf{T}} f(x) + z_0^{\mathsf{T}} h(x)}{y_0^{\mathsf{T}} g(x)}$$

    Let $k := 1$;

Step 1. Determine $(y_k, z_k) := \arg \max_{y \in \Sigma, z \geq 0} G(y, z, \lambda_{k-1})$;

Step 2. **If** $G(\lambda_{k-1}) = 0$

        Then (y_{k-1}, z_{k-1}) is an optimal solution of (D)

            with value λ_{k-1} and **Stop**.

        Else GoTo *Step* 3;

Step 3. Compute $\lambda_k := d(y_k, z_k)$, let $k := k + 1$, and **GoTo** *Step* 1.

Algorithm 4.3: "Dual" algorithm.

Lemma 4.2.1

For $(\widehat{y}, \widehat{z}) \in \Sigma \times I\!R_+^r$ we have

$$\mathcal{U}_d^{\circ}(d(\widehat{y}, \widehat{z})) = \left\{ (y, z) \in \Sigma \times I\!R_+^r : G(y, z, d(\widehat{y}, \widehat{z})) > 0 \right\}$$

and

$$\mathcal{U}_d(d(\widehat{y}, \widehat{z})) = \left\{ (y, z) \in \Sigma \times I\!R_+^r : G(y, z, d(\widehat{y}, \widehat{z})) \geq 0 \right\}.$$

Proof: We first consider the case that $\mathcal{U}_d^{\circ}(d(\widehat{y}, \widehat{z}))$ is nonempty, i.e. $(\widehat{y}, \widehat{z}) \in \Sigma \times I\!R_+^r$ is nonoptimal for (D). For $(y, z) \in \mathcal{U}_d^{\circ}(d(\widehat{y}, \widehat{z}))$ we have that $d(y, z) > d(\widehat{y}, \widehat{z})$. From Lemma 4.1.2 and $g(x) > 0$ for every $x \in S$ we see

$$G(y, z, d(\widehat{y}, \widehat{z})) = \min_{x \in S} \left\{ y^{\mathsf{T}} \left(f(x) - d(\widehat{y}, \widehat{z}) g(x) \right) + z^{\mathsf{T}} h(x) \right\}$$
$$> \min_{x \in S} \left\{ y^{\mathsf{T}} \left(f(x) - d(y, z) g(x) \right) + z^{\mathsf{T}} h(x) \right\} = 0.$$

Conversely, if $G(y, z, d(\widehat{y}, \widehat{z})) > 0$ and $(y, z) \in \Sigma \times I\!R_+^r$ then, using Lemma 4.1.2 it follows that $d(y, z) > d(\widehat{y}, \widehat{z})$ which concludes the proof for the nonempty case.

On the other hand, if $\mathcal{U}_d^{\circ}(d(\widehat{y}, \widehat{z}))$ is empty, then we know that $d(y, z) \leq d(\widehat{y}, \widehat{z})$ for every $(y, z) \in \Sigma \times I\!R_+^r$, and hence by Lemma 4.1.2 the set

$$\left\{ (y, z) \in \Sigma \times I\!R_+^r : G(y, z, d(\widehat{y}, \widehat{z})) > 0 \right\}$$

is also empty.

Finally the last equality can be proved in a similar way as the first part of this proof, and so we omit it. $\qquad\qquad\square$

In order to provide a clear geometric interpretation of the "dual" algorithm it is also important to analyze the lower approximation function $G_{(y,z)}$. Since this function is the minimum of a set of affine functions, it is concave and so by Corollary 10.1.1 of [117] it is continuous on $\mathbb{R}$. Moreover, alike the "primal" parametric function F, the function $G_{(y,z)}$ is decreasing, since $g(x) > 0$ for every $x \in \mathcal{S}$. This yields the geometrical interpretation of the "dual" algorithm in Figure 4.3.

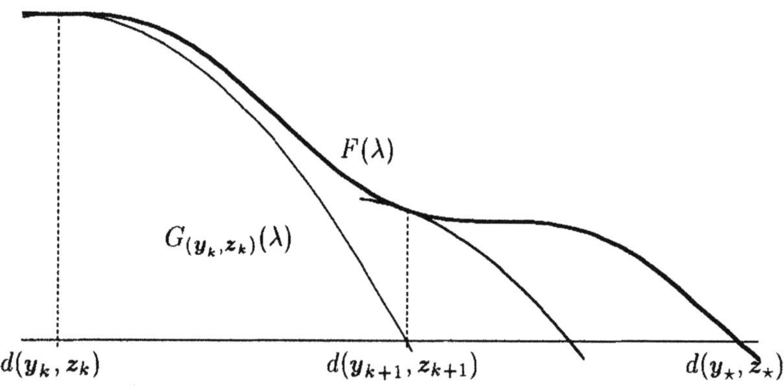

Figure 4.3: Geometric interpretation of the "dual" algorithm.

It is interesting to notice the similarities between this "dual" algorithm and the Dinkelbach-type algorithm, see Algorithm 4.1. In *Step* 1 a parametric problem must be solved to check in *Step* 2 whether or not optimality was reached. If the present iteration point is not optimal then the next iteration point is given by an optimal solution of the parametric problem solved in *Step* 1. Using this point, a "better" approximation of the optimal value is computed in *Step* 3. On the other hand, an essential difference between both algorithms is that while the above algorithm "walks" in the dual space, the Dinkelbach-type algorithm "moves" in the $\mathcal{X}$ space, yielding different approximation functions, see Figures 4.1 and 4.3. Accordingly, the Dinkelbach-type algorithm constructs a strictly decreasing sequence $\{\lambda_k : k \geq 1\}$ approaching the optimal value $\vartheta(P)$ from above, while the "dual" algorithm constructs a strictly increasing sequence $\{d(\boldsymbol{y}_k, \boldsymbol{z}_k) : k \geq 0\}$ approaching $\vartheta(P)$ from below.

To prove the convergence of the "dual" algorithm we need to investigate the behavior of the approximation function $G_{(\boldsymbol{y},\boldsymbol{z})}$. Observe that by Theorem 23.4 of [117] the subgradient set $\partial(-G_{(\boldsymbol{y},\boldsymbol{z})})(\lambda)$ of the convex function $-G_{(\boldsymbol{y},\boldsymbol{z})} : \mathbb{R} \longrightarrow \mathbb{R}$ at the point λ is nonempty. Recall that $\rho \in \mathbb{R}$ is a subgradient of the function $-G_{(\boldsymbol{y},\boldsymbol{z})}$ at the point λ if and only if

$$G_{(\boldsymbol{y},\boldsymbol{z})}(\lambda + t) \leq G_{(\boldsymbol{y},\boldsymbol{z})}(\lambda) - t\rho \qquad (4.10)$$

for every $t \in \mathbb{R}$. The next result characterizes the subgradient set $\partial(-G_{(\boldsymbol{y},\boldsymbol{z})})(\lambda)$. Before mentioning this result we introduce for fixed $(\boldsymbol{y}, \boldsymbol{z}) \in \Sigma \times \mathbb{R}^r_+$ the set $\mathcal{S}_{(\boldsymbol{y},\boldsymbol{z})}(\lambda)$ of optimal solutions of the optimization problem

$$\min_{\boldsymbol{x} \in \mathcal{S}} \left\{ \boldsymbol{y}^\top \boldsymbol{f}(\boldsymbol{x}) + \boldsymbol{z}^\top \boldsymbol{h}(\boldsymbol{x}) - \lambda \boldsymbol{y}^\top \boldsymbol{g}(\boldsymbol{x}) \right\}.$$

i.e.

$$\mathcal{S}_{(\boldsymbol{y},\boldsymbol{z})}(\lambda) := \left\{ \boldsymbol{x} \in \mathcal{S} : \boldsymbol{y}^\top \left(\boldsymbol{f}(\boldsymbol{x}) - \lambda \boldsymbol{g}(\boldsymbol{x}) \right) + \boldsymbol{z}^\top \boldsymbol{h}(\boldsymbol{x}) = G_{(\boldsymbol{y},\boldsymbol{z})}(\lambda) \right\}. \qquad (4.11)$$

Clearly, this set is nonempty. Also, by the continuity of the vector-valued functions $\boldsymbol{f}, \boldsymbol{g}$ and $\boldsymbol{h}$, it must be closed and thus, by the compactness of $\mathcal{S}$ and $\mathcal{S}_{(\boldsymbol{y},\boldsymbol{z})})(\lambda) \subseteq \mathcal{S}$, it is compact. Finally, if $\lambda \geq 0$ then the function $\boldsymbol{x} \longmapsto \boldsymbol{y}^\top \left(\boldsymbol{f}(\boldsymbol{x}) - \lambda \boldsymbol{g}(\boldsymbol{x}) \right) + \boldsymbol{z}^\top \boldsymbol{h}(\boldsymbol{x})$ is convex due to the convexity of $\boldsymbol{f}, \boldsymbol{h}$ and the concavity of $\boldsymbol{g}$, and this implies that $\mathcal{S}_{(\boldsymbol{y},\boldsymbol{z})}(\lambda) \subseteq \mathcal{S}$ is also convex for every $\lambda \geq 0$. Observe that the above result also holds for any λ if the functions g_j are affine for every $j \in J$.

Lemma 4.2.2

For every fixed $(\boldsymbol{y}, \boldsymbol{z}) \in \Sigma \times \mathbb{R}^r_+$ and $\lambda \in \mathbb{R}$ it follows that

$$\partial(-G_{(\boldsymbol{y},\boldsymbol{z})})(\lambda) = \left[\min_{\boldsymbol{x} \in \mathcal{S}_{(\boldsymbol{y},\boldsymbol{z})}(\lambda)} \left\{ \boldsymbol{y}^\top \boldsymbol{g}(\boldsymbol{x}) \right\}, \max_{\boldsymbol{x} \in \mathcal{S}_{(\boldsymbol{y},\boldsymbol{z})}(\lambda)} \left\{ \boldsymbol{y}^\top \boldsymbol{g}(\boldsymbol{x}) \right\} \right].$$

The proof of the above lemma is omitted since this result is a special case of a more general result given by Theorem 7.2 of [118] or Theorem 4.4.2 of [73]. However, an easy proof of this special case can be found in [14].

Denote now by $k^\star$ the number of times that *Step* 1 was executed by the algorithm. Clearly, if $k^\star$ is finite it follows that $G(y_{k^\star+1}, z_{k^\star+1}, d(y_{k^\star}, z_{k^\star})) = 0$, while for $k^\star = +\infty$ the algorithm does not stop. Before mentioning the next result we introduce

$$\Delta_k(y, z) := \max\left\{y^\top g(x) : x \in \mathcal{S}_{(y,z)}(d(y_k, z_k))\right\}$$

and

$$\delta_{k+1} := \min\left\{y_{k+1}^\top g(x) : x := \arg\min_{x \in \mathcal{S}} \frac{y_{k+1}^\top f(x) + z_{k+1}^\top h(x)}{y_{k+1}^\top g(x)}\right\}$$
$$= \min\left\{y_{k+1}^\top g(x) : x \in \mathcal{S}_{(y_{k+1}, z_{k+1})}(d(y_{k+1}, z_{k+1}))\right\}.$$

Observe that by Lemma 4.2.2 we have

$$\Delta_k(y, z) \in \partial(-G_{(y,z)})(d(y_k, z_k)) \text{ and } \delta_{k+1} \in \partial(-G_{(y_{k+1}, z_{k+1})})(d(y_{k+1}, z_{k+1})).$$

Theorem 4.2.1

The sequence $(y_k, z_k), 0 \leq k < k^\star$, does not contain optimal solutions of (D) and the corresponding function values $d(y_k, z_k), 0 \leq k < k^\star$ are strictly increasing. Moreover, if $k^\star$ is finite, then $d(y_{k^\star}, z_{k^\star}) = \lambda_\star = \vartheta(D)$ while for $k^\star = +\infty$ we have

$$\lim_{k \uparrow \infty} d(y_k, z_k) = \lambda_\star.$$

Finally, if $k^\star = +\infty$ and $(y_\star, z_\star)$ is an optimal solution of (D), then

$$\lambda_\star - d(y_{k+1}, z_{k+1}) \leq \left(1 - \frac{\Delta_k(y_\star, z_\star)}{\delta_{k+1}}\right)(\lambda_\star - d(y_k, z_k)) \tag{4.12}$$

holds for every $k \geq 0$.

Proof: Using Lemma 4.2.1 it follows that $G(y_{k+1}, z_{k+1}, d(y_k, z_k)) > 0$ if and only if (y_k, z_k) is nonoptimal. Moreover, by the same lemma we obtain that $d(y_{k+1}, z_{k+1}) > d(y_k, z_k)$ if (y_k, z_k) is nonoptimal, and so the first part of the theorem is proved.

Consider now the case with $k^\star$ finite. Since the algorithm stopped in a finite number of steps we must have $G(y_{k^\star+1}, z_{k^\star+1}, d(y_{k^\star}, z_{k^\star})) = 0$ and using again Lemma 4.2.1 it follows that $(y_{k^\star}, z_{k^\star})$ solves (D). Hence, we have $d(y_{k^\star}, z_{k^\star}) = \vartheta(P)$.

To verify the last part of the result, notice that $d(y_k, z_k)$, $k \geq 0$, is strictly increasing for $k^\star = +\infty$, and since $d(y_k, z_k) \leq \vartheta(D) < \infty$ for every $k \geq 0$, it must follow that $\lim_{k \uparrow \infty} d(y_k, z_k)$ exists and is finite-valued. Moreover, by Lemma 4.2.2 and (4.10) we obtain for every optimal solution $(y_\star, z_\star)$ of (D) that

$$G_{(y_\star, z_\star)}(d(y_\star, z_\star)) - G_{(y_\star, z_\star)}(d(y_k, z_k)) \leq -\left(d(y_\star, z_\star) - d(y_k, z_k)\right) \Delta_k(y_\star, z_\star).$$

Since $G_{(y_\star, z_\star)}(d(y_\star, z_\star)) = 0$ this implies that

$$\begin{aligned}
G_{(y_{k+1}, z_{k+1})}(d(y_k, z_k)) &= \max_{y \in \Sigma, z \geq 0} G_{(y,z)}(d(y_k, z_k)) \geq G_{(y_\star, z_\star)}(d(y_k, z_k)) \\
&\geq \left(d(y_\star, z_\star) - d(y_k, z_k)\right) \Delta_k(y_\star, z_\star).
\end{aligned} \tag{4.13}$$

On the other hand, applying again Lemma 4.2.2 and (4.10) we obtain

$$\begin{aligned}
G_{(y_{k+1}, z_{k+1})}(d(y_k, z_k)) &= G_{(y_{k+1}, z_{k+1})}(d(y_k, z_k)) - G_{(y_{k+1}, z_{k+1})}(d(y_{k+1}, z_{k+1})) \\
&\leq \left(d(y_{k+1}, z_{k+1}) - d(y_k, z_k)\right) \delta_{k+1}.
\end{aligned}$$

The above inequality and (4.13) imply that

$$\left(d(y_{k+1}, z_{k+1}) - d(y_k, z_k)\right) \delta_{k+1} \geq \left(d(y_\star, z_\star) - d(y_k, z_k)\right) \Delta_k(y_\star, z_\star). \tag{4.14}$$

Since $\Delta_k(y_\star, z_\star)$ and δ_{k+1} belong to the interval $[\delta, \Delta]$ with

$$\delta := \min_{x \in S} \min_{j \in J} g_j(x) > 0 \quad \text{and} \quad \Delta := \max_{x \in S} \max_{j \in J} g_j(x) < +\infty$$

it follows by (4.14) and the existence of $\lim_{k \uparrow \infty} d(y_k, z_k)$ that $\lim_{k \uparrow \infty} d(y_k, z_k) = d(y_\star, z_\star)$.

Finally, from (4.14) we obtain

$$\begin{aligned}
\vartheta(D) - d(y_{k+1}, z_{k+1}) &= \vartheta(D) - d(y_k, z_k) + d(y_k, z_k) - d(y_{k+1}, z_{k+1}) \\
&\leq \left(1 - \frac{\Delta_k(y_\star, z_\star)}{\delta_{k+1}}\right) (\vartheta(D) - d(y_k, z_k)).
\end{aligned}$$

$$\square$$

Clearly, by inequality (4.12) this algorithm converges at least linearly. In order to improve this convergence rate we need to impose a stronger constraint qualification than the Slater condition. Therefore, we will consider the following stronger Slater condition.

Strong Slater's condition

There exists some $\widehat{x} \in \mathrm{ri}(S)$ satisfying $h(\widehat{x}) < 0$.

This condition is by the continuity of the vector-valued function h equivalent to the requirement that there exists some $\widehat{x} \in S$ satisfying $h(\widehat{x}) < 0$. We can now establish the following simple condition which improves the convergence rate. However, before mentioning this result, we will show that the strong Slater condition implies the existence of an accumulation point of the sequence $\{(y_k, z_k)\}_{k \geq 0}$ generated by the dual algorithm. Observe that the existence of such an accumulation point is not immediately clear due to $z_k \geq 0$ for every $k \geq 0$.

Lemma 4.2.3

If the strong Slater condition holds, i.e. there exists some $\widehat{x} \in S$ satisfying $h(\widehat{x}) < 0$, then the sequence $\{(y_k, z_k)\}_{k \geq 0}$ has an accumulation point $(y_\star, z_\star)$ and this accumulation point is an optimal solution of (D). Moreover, if (D) has a unique optimal solution $(y_\star, z_\star)$ then $\lim_{k \uparrow \infty} y_k = y_\star$ and $\lim_{k \uparrow \infty} z_k = z_\star$.

Proof: Suppose that the sequence $(y_{k+1}, z_{k+1}) \in \Sigma \times I\!\!R_+^r$ has no convergent subsequence in $\Sigma \times I\!\!R_+^r$, $k \in K$. This implies, since Σ is a compact set, that there exists a subsequence $K_1 \subseteq K$ with

$$\lim_{k \in K_1, k \uparrow \infty} y_{k+1} = y_\star \quad \text{and} \quad \lim_{k \in K_1, k \uparrow \infty} \|z_{k+1}\| = \infty. \tag{4.15}$$

However,

$$d(y_k, z_k) = \min_{x \in S} \frac{y_k^\top f(x) + z_k^\top h(x)}{y_k^\top g(x)} \leq \frac{y_k^\top f(\widehat{x}) + z_k^\top h(\widehat{x})}{y_k^\top g(\widehat{x})}$$

and due to $g(\widehat{x}) > 0$, $h(\widehat{x}) < 0$ and (4.15) it follows that

$$\lim_{k \in K_1, k \uparrow \infty} d(y_{k+1}, z_{k+1}) = -\infty.$$

However, the sequence $d(y_{k+1}, z_{k+1})$ is strictly increasing by Theorem 4.2.1 and this yields a contradiction. Hence, the sequence $\{(y_k, z_k)\}_{k \geq 0}$ has an accumulation point $(\boldsymbol{y}_\star, \boldsymbol{z}_\star) \in \Sigma \times I\!R_+^r$. Applying again Theorem 4.2.1 and using the continuity of the function d we obtain that such an accumulation point $(\boldsymbol{y}_\star, \boldsymbol{z}_\star)$ is an optimal solution of (D), which concludes the proof the first part of this lemma.

The second part of this lemma is easily verified by contradiction. $\qquad\square$

Using the previous lemma we can establish the following improved convergence rate result.

Proposition 4.2.2

If the strong Slater condition holds and for every optimal solution $(\boldsymbol{y}_\star, \boldsymbol{z}_\star)$ of (D), the optimization problem

$$\min_{\boldsymbol{x} \in S} \frac{\boldsymbol{y}_\star^\top \boldsymbol{f}(\boldsymbol{x}) + \boldsymbol{z}_\star^\top \boldsymbol{h}(\boldsymbol{x})}{\boldsymbol{y}_\star^\top \boldsymbol{g}(\boldsymbol{x})} \qquad (D_\star)$$

has a unique optimal solution, then the "dual" algorithm converges superlinearly.

Proof: From Theorem 4.2.1 it follows that if for all optimal solutions $(\boldsymbol{y}_\star, \boldsymbol{z}_\star)$ of (D)

$$\limsup_{k \uparrow \infty} \left(1 - \frac{\Delta_k(\boldsymbol{y}_\star, \boldsymbol{z}_\star)}{\delta_{k+1}} \right) = 0$$

holds, then the convergence rate of Algorithm 4.3 is superlinear, and so the result is proved. Let $\delta_\infty := \limsup_{k \uparrow \infty} \delta_{k+1}$. By the definition of $\limsup$ there exists a subsequence $K \subseteq I\!N$ such that $\delta_\infty = \lim_{k \in K, k \uparrow \infty} \delta_{k+1}$. Moreover, by Lemma 4.2.3 it follows that the sequence $\{(\boldsymbol{y}_k, \boldsymbol{z}_k)\}_{k \geq 0}$ admits a subsequence $K_1 \subseteq K$ such that and $\lim_{k \in K_1, k \uparrow \infty}(\boldsymbol{y}_{k+1}, \boldsymbol{z}_{k+1}) = (\boldsymbol{y}_\star, \boldsymbol{z}_\star)$ with $(\boldsymbol{y}_\star, \boldsymbol{z}_\star)$ an accumulation point, which is an optimal solution of (D). Hence, consider the sequence

$$1 - \frac{\Delta_k(\boldsymbol{y}_\star, \boldsymbol{z}_\star)}{\delta_{k+1}}$$

for such an accumulation point $(\boldsymbol{y}_\star, \boldsymbol{z}_\star)$. It is easy to verify that the point-to-set mapping $(\boldsymbol{y}, \boldsymbol{z}) \longmapsto \partial(-G_{(\boldsymbol{y},\boldsymbol{z})})(d(\boldsymbol{y}, \boldsymbol{z}))$ is upper semicontinuous. Since

$$\delta_{k+1} \in \partial(-G_{(\boldsymbol{y}_{k+1}, \boldsymbol{z}_{k+1})})(d(\boldsymbol{y}_{k+1}, \boldsymbol{z}_{k+1})), \quad \lim_{k \in K_1, k \uparrow \infty} \delta_{k+1} = \delta_\infty \text{ and}$$

$$\lim_{k \in K_1, k \uparrow \infty} (\boldsymbol{y}_{k+1}, \boldsymbol{z}_{k+1}) = (\boldsymbol{y}_\star, \boldsymbol{z}_\star)$$

we obtain that

$$\delta_\infty \in \partial(-G_{(y_\star,z_\star)})(d(y_\star,z_\star)). \tag{4.16}$$

On the other hand, it is clear by Lemma 4.2.2 that

$$\Delta_k(y_\star,z_\star) \in \partial\left(-G_{(y_\star,z_\star)}\right)(d(y_k,z_k)).$$

Moreover, since the increasing sequence $d(y_k,z_k)$ converges from below to $d(y_\star,z_\star)$, it follows by the convexity of the function $-G_{(y_\star,z_\star)}$ and

$$\Delta_k(y_\star,z_\star) \in \partial(-G_{(y_\star,z_\star)})(d(y_k,z_k))$$

that

$$\Delta_k(y_\star,z_\star) \leq \Delta_{k+1}(y_\star,z_\star) \leq \cdots \leq a_\star \text{ with } a_\star \in \partial(-G_{(y_\star,z_\star)})(d(y_\star,z_\star)).$$

This implies that $\lim_{k\uparrow\infty} \Delta_k(y_\star,z_\star) =: \Delta_\infty(y_\star,z_\star)$ exists and by the upper semi-continuity of the point-to-set mapping $\lambda \longmapsto \partial(-G_{(y_\star,z_\star)})(\lambda)$ we obtain that

$$\Delta_\infty(y_\star,z_\star) \in \partial(-G_{(y_\star,z_\star)})(d(y_\star,z_\star)).$$

Since we already observed that

$$\Delta_\infty(y_\star,z_\star) \leq a_\star \text{ for every } a_\star \in \partial(-G_{(y_\star,z_\star)})(d(y_\star,z_\star)),$$

it must follow by Lemma 4.2.2 that

$$\Delta_\infty(y_\star,z_\star) = \min\left\{ y_\star^\mathsf{T} g(x) : x \in \mathcal{S}_{(y_\star,z_\star)}(d(y_\star,z_\star)) \right\}. \tag{4.17}$$

Observe now that by (4.16) and (4.17) we have

$$0 \leq \limsup_{k\uparrow\infty}\left(1 - \frac{\Delta_k(y_\star,z_\star)}{\delta_{k+1}}\right) = 1 - \liminf_{k\uparrow\infty}\frac{\Delta_k(y_\star,z_\star)}{\delta_{k+1}} = 1 - \frac{\Delta_\infty(y_\star,z_\star)}{\delta_\infty} < 1.$$

Clearly, by (4.16), (4.17) and Lemma 4.2.2 the above limsup equals zero if the optimization problem $(D_\star)$ has a unique optimal solution. $\qquad\square$

In order to guarantee the uniqueness condition expressed in the above proposition we need to introduce the following subset of quasiconvex functions.

Definition 4.2.1 ([6])

The function $\psi : C \longrightarrow \mathbb{R}$ is called strictly quasiconvex if for each $x_1, x_2 \in C$ with $x_1 \neq x_2$

$$\psi(\lambda x_1 + (1 - \lambda)x_2) < \max\{\psi(x_1), \psi(x_2)\}$$

for every $0 < \lambda < 1$.

Observe by Proposition 3.29 of [6] that $\min_{x \in \mathcal{X}} \psi(x)$ has a *unique* optimal solution if the continuous function $\psi : C \longrightarrow \mathbb{R}$ is strictly quasiconvex on $\mathcal{X}$, with $\mathcal{X} \subset C$. The next corollary establishes sufficient conditions on the functions f_j and g_j to ensure that the convergence rate of the "dual" algorithm is superlinear.

Corollary 4.2.1

If the strong Slater condition holds and either one of the following conditions

(i) $f : C \longrightarrow \mathbb{R}^n$ is on $\mathcal{S}$ and nonnegative on $\mathcal{X}$ and $g : C \longrightarrow \mathbb{R}^n$ is positive and concave on $\mathcal{S}$;

(ii) $f : C \longrightarrow \mathbb{R}^n$ is convex on $\mathcal{S}$ and nonnegative on $\mathcal{X}$ and $g : C \longrightarrow \mathbb{R}^n$ is positive and strictly concave on $\mathcal{S}$;

(iii) $f : C \longrightarrow \mathbb{R}^n$ is strictly convex on $\mathcal{S}$ and $g : C \longrightarrow \mathbb{R}^n$ is positive and affine on $\mathcal{S}$

then the "dual" algorithm converges superlinearly.

Proof: We will prove the result only for *(i)* and *(ii)*. For *(iii)* the result follows easily. By Proposition 4.2.2 and the properties of strictly quasiconvex functions it is sufficient to show that for any optimal solution $(y_\star, z_\star)$ of (D) the function $\psi : C \longrightarrow \mathbb{R}$ given by

$$\psi(x) := \frac{y_\star^\top f(x) + z_\star^\top h(x)}{y_\star^\top g(x)}$$

is strictly quasiconvex on $\mathcal{S}$.

By Proposition 4.2.1 we have that

$$\min_{x \in \mathcal{S}} \frac{y_\star^\top f(x) + z_\star^\top h(x)}{y_\star^\top g(x)}$$

equals $\vartheta(P) \geq 0$ and so it follows that

$$x \longmapsto y_\star^\mathsf{T} f(x) + z_\star^\mathsf{T} h(x)$$

is nonnegative on $\mathcal{S}$. This yields for every $x_1, x_2 \in \mathcal{S}$ with $x_1 \neq x_2$ and $0 < \lambda < 1$ that

$$
\begin{aligned}
&\psi(\lambda x_1 + (1-\lambda)x_2) \\
&< \frac{\lambda y_\star^\mathsf{T} f(x_1) + \lambda z_\star^\mathsf{T} h(x_1) + (1-\lambda)y_\star^\mathsf{T} f(x_2) + (1-\lambda)z_\star^\mathsf{T} h(x_2)}{\lambda y_\star^\mathsf{T} g(x_1) + (1-\lambda)y_\star^\mathsf{T} g(x_2)} \\
&= \frac{\lambda y_\star^\mathsf{T} g(x_1)\dfrac{y_\star^\mathsf{T} f(x_1) + z_\star^\mathsf{T} h(x_1)}{y_\star^\mathsf{T} g(x_1)} + (1-\lambda)y_\star^\mathsf{T} g(x_2)\dfrac{y_\star^\mathsf{T} f(x_2) + z_\star^\mathsf{T} h(x_2)}{y_\star^\mathsf{T} g(x_2)}}{\lambda y_\star^\mathsf{T} g(x_1) + (1-\lambda)y_\star^\mathsf{T} g(x_2)} \\
&\leq \max\left\{\frac{y_\star^\mathsf{T} f(x_1) + z_\star^\mathsf{T} h(x_1)}{y_\star^\mathsf{T} g(x_1)}, \frac{y_\star^\mathsf{T} f(x_2) + z_\star^\mathsf{T} h(x_2)}{y_\star^\mathsf{T} g(x_2)}\right\} \\
&= \max\{\psi(x_1), \psi(x_2)\}
\end{aligned}
$$

which completes the proof. $\qquad\square$

The practicality of this algorithm depends mostly on how *Step* 1 is solved. Unfortunately, solving (D_λ) takes a lot of time, and this will influence the practical applicability of the new method. On the other hand, when applying the Dinkelbach-type algorithm we need to solve in each step the optimization problem (P_λ) which appears to be easier. Moreover, as observed (D_λ) corresponds to a Lagrangian dual of (P_λ) and thus we can relate an optimal solution x_{k+1} of (P_{λ_k}) to an optimal solution (y_{k+1}, z_{k+1}) of (D_{λ_k}). To derive this relation, we will assume additionally that f_j, g_j, $j \in J$ and $h_l : \mathbb{R}^m \longrightarrow \mathbb{R}$, $l = 1, \ldots, r$, are also differentiable functions and that the nonempty compact convex set $\mathcal{S}$ is given by

$$\mathcal{S} := \{x \in \mathbb{R}^m : p_l(x) \leq 0, l = 1, \ldots, s\}$$

where $p_l : \mathbb{R}^m \longrightarrow \mathbb{R}$, $l = 1, \ldots, s$, are convex and differentiable functions.

Clearly, (P_{λ_k}) is equivalent to the following convex programming problem

$$
\begin{aligned}
\min \quad & t \\
\text{s.t.:} \quad & q_j(x) - t \leq 0 \quad \forall j = 1, \ldots, n
\end{aligned}
$$

$$h_l(\boldsymbol{x}) \leq 0 \qquad \forall l = 1, \ldots, r$$

$$p_l(\boldsymbol{x}) \leq 0 \qquad \forall l = 1, \ldots, s$$

with $q_j(\boldsymbol{x}) := f_j(\boldsymbol{x}) - \lambda_k g_j(\boldsymbol{x})$, $j = 1, \ldots, n$. Let $\boldsymbol{x}_{k+1}$ and t_{k+1} be an optimal solution of the above problem, and define $J' := \{1 \leq j \leq n : q_j(\boldsymbol{x}_{k+1}) = t_{k+1}\}$, $R' := \{1 \leq l \leq r : h_l(\boldsymbol{x}_{k+1}) = 0\}$ and $S' := \{1 \leq l \leq s : p_l(\boldsymbol{x}_{k+1}) = 0\}$. Since the Slater condition holds the Karush-Kuhn-Tucker conditions ensure the existence of nonnegative scalars $u_j, j \in J'$, $v_l, l \in R'$ and $\xi_l, l \in S'$ satisfying

$$\sum_{j \in J'} u_j \nabla q_j(\boldsymbol{x}_{k+1}) + \sum_{l \in R'} v_l \nabla h_l(\boldsymbol{x}_{k+1}) + \sum_{l \in S'} \xi_l \nabla p_l(\boldsymbol{x}_{k+1}) = 0 \qquad (4.18)$$

$$\sum_{j \in J'} u_j = 1 \qquad (4.19)$$

$$(\boldsymbol{u}_{J'}, \boldsymbol{v}_{R'}, \boldsymbol{\xi}_{S'}) \geq 0. \qquad (4.20)$$

Notice that the set J' is nonempty, due to the optimality of $(\boldsymbol{x}_{k+1}, t_{k+1})$. It is now possible to relate the scalars u_j, $j \in J'$ and v_l, $l \in R'$ to an optimal solution of (D_{λ_k}).

Lemma 4.2.4

An optimal solution $(\widehat{\boldsymbol{y}}, \widehat{\boldsymbol{z}})$ of (D_{λ_k}) is given by

$$\widehat{y}_j = \begin{cases} 0 & \text{if } j \notin J' \\ u_j & \text{if } j \in J' \end{cases}, \qquad \widehat{z}_l = \begin{cases} 0 & \text{if } l \notin R' \\ v_l & \text{if } l \in R' \end{cases}$$

where $\boldsymbol{u}_{J'}, \boldsymbol{v}_{R'}$ solve the system (4.18), (4.19), (4.20).

Proof: From (4.19) and (4.20) it follows that $\widehat{\boldsymbol{y}}$ belongs to Σ and that $\widehat{\boldsymbol{z}} \geq 0$. Moreover, by the definition of J' and R' we obtain that

$$\sum_{j \in J'} \widehat{y}_j q_j(\boldsymbol{x}_{k+1}) + \sum_{l \in R'} \widehat{z}_l h_l(\boldsymbol{x}_{k+1}) = t_{k+1}.$$

Hence, since (D_{λ_k}) is the Lagrangian dual of (P_{λ_k}) it follows using the previous equality that

$$\sum_{j \in J'} \widehat{y}_j q_j(\boldsymbol{x}_{k+1}) + \sum_{l \in R'} \widehat{z}_l h_l(\boldsymbol{x}_{k+1}) = \min_{\boldsymbol{x} \in \mathcal{X}} \max_{j \in J} \{f_j(\boldsymbol{x}) - \lambda_k g_j(\boldsymbol{x})\}$$

$$= \max_{\boldsymbol{y} \in \Sigma, \boldsymbol{z} \geq 0} G(\boldsymbol{y}, \boldsymbol{z}, \lambda_k).$$

It is left to show that the pair $(\widehat{y}, \widehat{z})$ is an optimal solution of

$$\max_{y \in \Sigma, z \geq 0} G(y, z, \lambda_k).$$

Since $\min_{x \in S}\{\widehat{y}^\mathsf{T} q(x) + \widehat{z}^\mathsf{T} h(x)\}$ is a convex optimization problem, the Karush-Kuhn-Tucker conditions are sufficient, see [73]. Clearly, by the definition of $(\widehat{y}, \widehat{z})$ and (4.18), (4.19), (4.20) the vector x_{k+1} satisfies these conditions, and thus x_{k+1} is an optimal solution of $\min_{x \in S}\{\widehat{y}^\mathsf{T} q(x) + \widehat{z}^\mathsf{T} h(x)\}$. Hence, $(\widehat{y}, \widehat{z}) \in \Sigma \times I\!\!R^r_+$ satisfy

$$\max_{y \in \Sigma, z \geq 0} G(y, z, \lambda_k) = \sum_{j \in J'} \widehat{y}_j q_j(x_{k+1}) + \sum_{l \in R'} \widehat{z}_l h_l(x_{k+1}) = \min_{x \in S}\{\widehat{y}^\mathsf{T} q(x) + \widehat{z}^\mathsf{T} h(x)\}$$
$$= G(\widehat{y}, \widehat{z}, \lambda_k)$$

and so $(\widehat{y}, \widehat{z})$ solves (D_{λ_k}). $\qquad\qquad\qquad\qquad\qquad\qquad\qquad\qquad\qquad\square$

This lemma provides an easy procedure to solve *Step* 1, and thus to obtain the next iteration (y_{k+1}, z_{k+1}) of the Algorithm 4.3. However, due to numerical errors the Karush-Kuhn-Tucker system may appear to be "inconsistent". To solve this problem, observe first that the linear system (4.18), (4.19), (4.20) can be rewritten as follows

$$Au + B_1 v + B_2 \xi = 0, u \in \Sigma, v, \xi \geq 0.$$

Letting $E := [AB_1 B_2]^\mathsf{T}[AB_1 B_2]$ and $w = (u, v, \xi)$ it follows that solving this linear system corresponds to finding a nonnegative vector $w \in I\!\!R^\eta$ where $\eta := |J'| + |R'| + |S'|$, with the smallest ellipsoidal norm $\sqrt{w^\mathsf{T} E w}$, under the constraint that its first $|J'|$ components belong to the unit simplex, or equivalently

$$\min \frac{1}{2} w^\mathsf{T} E w \tag{4.21}$$
$$u \in \Sigma, v, \xi \geq 0. \tag{4.22}$$

Clearly, in the presence of no numerical errors the optimal value of this problem is zero.

In order to conclude the discussion of the "dual" algorithm it is important to consider a stopping rule for *Step* 2. Due to Proposition 4.2.1, the stopping rule can be derived similarly as for the Dinkelbach-type algorithm, see Section 4.1.1. In fact, from (4.5) of Proposition 4.1.1 it follows that stopping the "dual" algorithm whenever

$F(\lambda_k) \leq \varepsilon \underline{g}(\boldsymbol{x}_k)$, with $\lambda_k = d(\boldsymbol{y}_k, \boldsymbol{z}_k)$ the current iteration point and $\boldsymbol{x}_k \in \mathcal{X}$ an optimal solution of (P_{λ_k}), yields $\vartheta(P) - \lambda_k \leq \varepsilon$. Clearly, the stopping rule derived for the Dinkelbach-type algorithm (4.3) can also be used.

Apparently the above stopping rule ensures only that the lower approximation λ_k belongs to a ε-neighborhood of the optimal value. Since the "dual" algorithm also provides a feasible primal solution $\boldsymbol{x}_k$ it is important to analyze its "quality". Using $\lambda_k \leq \lambda_\star = \vartheta(P)$ we have for $\overline{\lambda} := \max_{j \in J} \dfrac{f_j(\boldsymbol{x}_k)}{g_j(\boldsymbol{x}_k)}$ that

$$
\begin{aligned}
F(\lambda_k) &= \max_{j \in J} \left\{ f_j(\boldsymbol{x}_k) - \lambda_k g_j(\boldsymbol{x}_k) \right\} \\
&\geq \max_{j \in J} \left\{ f_j(\boldsymbol{x}_k) - \lambda_\star g_j(\boldsymbol{x}_k) \right\} = \max_{j \in J} \left\{ f_j(\boldsymbol{x}_k) - \overline{\lambda} g_j(\boldsymbol{x}_k) + (\overline{\lambda} - \lambda_\star) g_j(\boldsymbol{x}_k) \right\} \\
&\geq 0 + (\overline{\lambda} - \lambda_\star)\, g_{j_k}(\boldsymbol{x}_k)
\end{aligned}
$$

with $j_k := \arg \max_{j \in J} \dfrac{f_j(\boldsymbol{x}_k)}{g_j(\boldsymbol{x}_k)}$. From the above inequality it follows that

$$
(\overline{\lambda} - \lambda_\star) \leq \frac{F(\lambda_k)}{\underline{g}(\boldsymbol{x}_k)} \leq \varepsilon.
$$

Therefore, the described stopping rule ensures that both the current dual solution value and the primal solution value belong to a ε-neighborhood.

Similar to the Dinkelbach type-2 approach it is possible to construct a scaled version the "dual" algorithm. However, the derivation of this scaled version is quite technical, see [15], and produces the same type of convergence rate results. Moreover, contrary to the Dinkelbach type-2 version, the scaled version does not improve the performance of its original version, see [15].

We will now specialize the derived results to the linear case.

Linear case

Let the functions f_j, g_j, for every $j \in J$ and the feasible set be given by

$$
f_j(\boldsymbol{x}) := \boldsymbol{a}_{.j}^{\mathsf{T}} \boldsymbol{x} + \alpha_j, \quad g_j(\boldsymbol{x}) := \boldsymbol{b}_{.j}^{\mathsf{T}} \boldsymbol{x} + \beta_j \quad \text{and} \quad \mathcal{X} := \left\{ \boldsymbol{x} \in \mathbb{R}^m : \boldsymbol{C}^{\mathsf{T}} \boldsymbol{x} \leq \boldsymbol{\gamma}, \boldsymbol{x} \geq 0 \right\}
$$

where $\boldsymbol{a}_{.j}$, $\boldsymbol{b}_{.j}$ denote respectively the jth column of the $m \times n$ matrix $\boldsymbol{A}$ and $\boldsymbol{B}$, $\boldsymbol{\alpha}^{\mathsf{T}} = [\alpha_1, \ldots, \alpha_n]$, $\boldsymbol{\beta}^{\mathsf{T}} = [\beta_1, \ldots, \beta_n]$, $\boldsymbol{C}$ a $m \times r$ matrix and $\boldsymbol{\gamma} \in \mathbb{R}^r$. We will also assume that

(A_1) $\mathcal{X} \subset \mathbb{R}^m$ is nonempty and bounded;

(A_2) $\boldsymbol{B}^\mathsf{T}\boldsymbol{x} + \beta > 0$ for all $\boldsymbol{x} \in \mathcal{X}$.

Thus, our generalized linear fractional programming problem is given by

$$\min_{\boldsymbol{x}\in\mathcal{X}} \left\{ \max_{j\in J} \frac{\boldsymbol{a}_{.j}^\mathsf{T}\boldsymbol{x} + \alpha_j}{\boldsymbol{b}_{.j}^\mathsf{T}\boldsymbol{x} + \beta_j} \right\}. \qquad (LP)$$

Consider also the following optimization problem problem

$$\min_{(\boldsymbol{x},x_0)\in\mathcal{X}_0} \left\{ \max_{j\in J} \frac{\boldsymbol{a}_{.j}^\mathsf{T}\boldsymbol{x} + \alpha_j x_0}{\boldsymbol{b}_{.j}^\mathsf{T}\boldsymbol{x} + \beta_j x_0} \right\} \qquad (LP_0)$$

with $\mathcal{X}_0 := \left\{ (\boldsymbol{x}, x_0) \in \mathbb{R}^{m+1}, \boldsymbol{C}^\mathsf{T}\boldsymbol{x} \leq \gamma x_0, \sum_{i=1}^{m} x_i + x_0 = 1, \boldsymbol{x} \geq 0, x_0 \geq 0 \right\}$. Before analyzing the relation between the above optimization problem and (LP), we will recall the definition of equivalent problems.

Definition 4.2.2 ([40])

The two optimization problems

$$\max\{F(\boldsymbol{x}) : \boldsymbol{x} \in \mathcal{X}\} \;\; and \;\; \max\{G(\boldsymbol{x}) : \boldsymbol{x} \in \mathcal{Y}\}$$

are called equivalent if there exists a one-to-one mapping ϕ of the feasible set $\mathcal{X}$ onto $\mathcal{Y}$ such that $F(\boldsymbol{x}) := G(\phi(\boldsymbol{x}))$ for each $\boldsymbol{x} \in \mathcal{X}$.

It is now possible to relate the two optimization problems (LP) and (LP_0).

Lemma 4.2.5

If (A_1) holds then (LP) and (LP_0) are equivalent problems.

Proof: In order to exhibit a one-to-one mapping of $\mathcal{X}$ onto $\mathcal{X}_0$ we will first show that for any $(\boldsymbol{x}, x_0) \in \mathcal{X}_0$, the scalar x_0 can never be zero. Suppose that there exists some $(\boldsymbol{x}, x_0) \in \mathcal{X}_0$ such that x_0 equals 0. Thus, $\boldsymbol{C}^\mathsf{T}\boldsymbol{x} \leq 0$ and $\sum_{i=1}^{m} x_j = 1$ with $x_i \geq 0$, $i = 1,\ldots,m$. It follows now for any $\boldsymbol{w} \in \mathcal{X}$ and $t > 0$ that

$$\boldsymbol{w} + t\boldsymbol{x} \geq 0 \;\; and \;\; \boldsymbol{C}^\mathsf{T}\boldsymbol{w} + t\boldsymbol{C}^\mathsf{T}\boldsymbol{x} \leq \gamma.$$

Hence, $\boldsymbol{w} + t\boldsymbol{x} \in \mathcal{X}$ for all $t > 0$ which contradicts the assumption (A_1) that $\mathcal{X}$ is a bounded set. Consider the mapping ϕ of $\mathcal{X}$ into $\mathcal{X}_0$ given by $\phi(\boldsymbol{x}) := \frac{1}{1+\sum_{i=1}^{m} x_i}(\boldsymbol{x}, 1)$. Clearly, the image of $\mathcal{X}$ under ϕ is contained in $\mathcal{X}_0$. Moreover, for all $(\boldsymbol{x}, x_0) \in \mathcal{X}_0$ there exists a unique point in $\mathcal{X}$ given by $\frac{\boldsymbol{x}}{x_0}$ and thus ϕ is a one-to-one mapping of $\mathcal{X}$ onto $\mathcal{X}_0$. Also, it follows easily that the objective value of (LP) at $\boldsymbol{x} \in \mathcal{X}$ equals the objective value of (LP_0) at $\phi(\boldsymbol{x})$ and so this concludes the proof. $\square$

From the above lemma it follows that the denominators of (LP_0) are always positive for $(\boldsymbol{x}, x_0) \in \mathcal{X}_0$. On the other hand, by assumption (A_1) we obtain that $\mathcal{X}_0$ is a nonempty bounded set. Therefore, if assumptions (A_1) and (A_2) hold then the optimization problem (LP_0) corresponds to a standard generalized linear fractional problem.

Observe also that the mapping used in the proof of Lemma 4.2.5 is equivalent to the Charnes and Cooper transformation [34], see Section 3.1. Therefore, Lemma 4.2.5 is comparable to the results derived by Charnes and Cooper [34] and in particular to the proof of equivalence between the different problems. However, while the Charnes and Cooper transformation is used to reduce a single-ratio fractional linear programming problem into a linear programming problem, the transformation used in the context of Lemma 4.2.5 maintains the structure of the original problem, see Section 3.1.

The transformation of (LP) into (LP_0) will enable us to apply directly the results derived for the nonlinear case to (LP_0). In fact, the feasible set of (LP_0) can be decomposed into the convex cone $\{(\boldsymbol{x}, x_0) \in \mathbb{R}^{m+1} : \boldsymbol{C}^\top \boldsymbol{x} \leq \gamma x_0\}$ and the compact convex set $\mathcal{S}_0 := \{(\boldsymbol{x}, x_0) \in \mathbb{R}^{m+1} : x_0 + \sum_{i=1}^{m} x_i = 1, \boldsymbol{x} \geq 0, x_0 \geq 0\}$. In order to derive the dual problem of (LP_0), the additional assumption $\boldsymbol{B}^\top \boldsymbol{x} + \beta x_0 > 0$ for all $(\boldsymbol{x}, x_0) \in \mathcal{S}_0$ is required. Observe this is guaranteed by the stronger positivity assumption (A_2') $\boldsymbol{B} > 0, \beta > 0$. We can now state the dual of (LP_0)

$$\max_{\boldsymbol{y} \in \Sigma, \boldsymbol{z} \geq 0} \min_{(\boldsymbol{x}, x_0) \in \mathcal{S}_0} \frac{\boldsymbol{y}^\top(\boldsymbol{A}^\top \boldsymbol{x} + \alpha x_0) + \boldsymbol{z}^\top(\boldsymbol{C}^\top \boldsymbol{x} - \gamma x_0)}{\boldsymbol{y}^\top(\boldsymbol{B}^\top \boldsymbol{x} + \beta x_0)} \qquad (D_0)$$

formed by the constraints related to the original problem. The new algorithm described in the previous section constructs a sequence $(\boldsymbol{y}_k, \boldsymbol{z}_k) \in \Sigma \times \mathbb{R}_+^r$ with function values $d(\boldsymbol{y}_k, \boldsymbol{z}_k)$ approximating from below the optimal value of (LP_0). Recall that by Lemma 4.2.5, the value $\vartheta(P_0)$ equals $\vartheta(P)$. Hence, for a given λ the new algorithm solves in *Step* 1 the parametric problem (D_{0_λ})

$$\max_{\boldsymbol{y}\in\Sigma,\boldsymbol{z}\geq 0}\left\{\min_{(\boldsymbol{x},x_0)\in\mathcal{S}_0}\left\{\left(\boldsymbol{y}^\top\left(\boldsymbol{A}-\lambda\boldsymbol{B}\right)^\top+\boldsymbol{z}^\top\boldsymbol{C}^\top\right)\boldsymbol{x}+\left(\boldsymbol{y}^\top\left(\boldsymbol{\alpha}-\lambda\boldsymbol{\beta}\right)-\boldsymbol{z}^\top\boldsymbol{\gamma}\right)x_0\right\}\right\}.$$

The next iteration point, $(\boldsymbol{y},\boldsymbol{z})$, is given by an optimal solution of the above problem. It is left to evaluate the value of the objective function d of (D_0) at this point, i.e. to compute $d(\boldsymbol{y},\boldsymbol{z})$. In this case, this corresponds to solving the following linear fractional programming problem

$$\min_{(\boldsymbol{x},x_0)\in\mathcal{S}_0}\frac{\boldsymbol{y}^\top(\boldsymbol{A}^\top\boldsymbol{x}+\boldsymbol{\alpha}x_0)+\boldsymbol{z}^\top(\boldsymbol{C}^\top\boldsymbol{x}-\boldsymbol{\gamma}x_0)}{\boldsymbol{y}^\top(\boldsymbol{B}^\top\boldsymbol{x}+\boldsymbol{\beta}x_0)}. \tag{4.23}$$

Observe that the objective function in (4.23) is a ratio of linear functions, and thus quasiconcave. Since a quasiconcave function attains its minimum over a compact convex set at an extreme point (see [6]), it follows that the optimal value of (4.23) has the following special form

$$d(\boldsymbol{y},\boldsymbol{z})=\min\left\{\frac{\boldsymbol{\alpha}^\top\boldsymbol{y}-\boldsymbol{\gamma}^\top\boldsymbol{z}}{\boldsymbol{\beta}^\top\boldsymbol{y}},\min_{1\leq i\leq m}\frac{\boldsymbol{a}_{i.}\boldsymbol{y}+\boldsymbol{c}_{i.}\boldsymbol{z}}{\boldsymbol{b}_{i.}\boldsymbol{y}}\right\}, \tag{4.24}$$

where $\boldsymbol{a}_{i.}$, $\boldsymbol{b}_{i.}$ and $\boldsymbol{c}_{i.}$ denote respectively the ith line of $\boldsymbol{A},\boldsymbol{B}$ and $\boldsymbol{C}$. This observation implies that (D_0) corresponds to the following generalized linear fractional programming problem

$$\max_{\boldsymbol{y}\in\Sigma,\boldsymbol{z}\geq 0}\left\{\min\left\{\frac{\boldsymbol{\alpha}^\top\boldsymbol{y}-\boldsymbol{\gamma}^\top\boldsymbol{z}}{\boldsymbol{\beta}^\top\boldsymbol{y}},\min_{1\leq i\leq m}\frac{\boldsymbol{a}_{i.}\boldsymbol{y}+\boldsymbol{c}_{i.}\boldsymbol{z}}{\boldsymbol{b}_{i.}\boldsymbol{y}}\right\}\right\} \tag{LD}$$

which is the standard dual problem of a generalized linear fractional program, described in [42, 43, 79], under assumption (A_2'). Observed that the above standard dual problem of (P) can be derived using a weaker assumption than (A_1). In fact, in [42, 79] instead of (A_1) it is only required that the feasible set should be nonempty.

Crouzeix *et al.* discuss in [43] how to solve (LP) whenever the feasible set $\mathcal{X}$ is not bounded. In this case, they show that the Dinkelbach-type algorithm applied to the standard dual problem (LD) converges and recovers the optimal solution value. Therefore, it is appropriate to relate this approach to our Algorithm 4.3. Observe that the Dinkelbach-type algorithm applied to (LD) requires solving the following parametric problem for a given λ

$$\max_{\boldsymbol{y}\in\Sigma,\boldsymbol{z}\geq 0}\min\left\{(\boldsymbol{\alpha}-\lambda\boldsymbol{\beta})^\top\boldsymbol{y}-\boldsymbol{\gamma}^\top\boldsymbol{z},\min_{1\leq i\leq m}\{(\boldsymbol{a}_{i.}-\lambda\boldsymbol{b}_{i.})\boldsymbol{y}+\boldsymbol{c}_{i.}\boldsymbol{z}\}\right\}. \tag{LD_λ}$$

However, due to the special form of (4.23), it follows that the above parametric problem corresponds to (D_{0_λ}). Also, in Dinkelbach's algorithm the next iteration value is given by (4.24) and hence the two algorithms are identical. Therefore, the Algorithm 4.3 extends to the nonlinear case the Dinkelbach-type algorithm applied to the dual of a generalized linear fractional program. Nevertheless, it is important to stress that in order to apply the "dual" algorithm it is required that the feasible set $\mathcal{X}$ is compact.

Since for (LP) the corresponding set $\mathcal{S}$ would be given by the noncompact set $\mathbb{R}^m_+$, while for (LP_0) the associated $\mathcal{S}_0$ is compact, it follows that by considering (LP_0) instead of (LP) the results derived in the previous section can smoothly be applied to the linear case. Observe also that by specializing Proposition 4.2.2 we retrieve a sufficient condition to guarantee superlinear convergence for the Dinkelbach-type algorithm applied to a generalized linear fractional program. Due to the special form of (4.23), it follows that (4.23) has a unique solution if

$$\min\left\{\frac{\alpha^\top y_\star - \gamma^\top z_\star}{\beta^\top y_\star}, \min_{1\leq i\leq m}\frac{a_{i.}y_\star + c_{i.}z_\star}{b_{i.}y_\star}\right\}$$

is uniquely attained. Therefore, the sufficient condition demands that for each optimal solution of (LD) only one ratio is active. Observe this implies that at a neighborhood of the optimal point the associated parametric function is concave, see Proposition 4.1 of [43]. Hence, in the neighborhood of the optimum, the Dinkelbach-type algorithm "coincides" with Newton's algorithm, and thus its convergence rate is superlinear.

4.2.2   A New Duality Approach

In this section we will consider the following generalized fractional programming problem (P) defined on the feasible set $\mathcal{X}$ given by $\mathcal{X} := \{x \in \mathcal{S} : h(x) \leq 0\}$, with $\mathcal{S} \subset R^m$ some compact convex set and $h : \mathbb{R}^m \longrightarrow \mathbb{R}^r$ a vector-valued convex function. We will also consider that the class of continuous functions $f_j, g_j : \mathcal{C} \longrightarrow \mathbb{R}$, $j \in J$, with $\mathcal{S} \subset \mathcal{C}$ and $\mathcal{C}$ an open set, are respectively convex and concave on the compact convex set $\mathcal{X}$. In addition, it is assumed that g_j are positive on $\mathcal{X}$ and that either the functions f_j are nonnegative on $\mathcal{X}$ or the functions g_j are affine for every $j \in J$. Observe that the above convexity requirements are weaker than the ones needed in the previous section. Accordingly, this class includes convex generalized fractional programming as a special case. For $f(x)^\top := (f_1(x), \ldots, f_n(x))$ and $g(x)^\top := (g_1(x), \ldots, g_n(x))$, it follows by the quasiconvexity of the function $\varphi : \mathbb{R} \times \mathbb{R}_+ \longrightarrow \mathbb{R}$ given by $\varphi(z) := \frac{z_1}{z_2}$ that

$$\max_{j \in J} \frac{f_j(x)}{g_j(x)} = \max_{y \in \Sigma} \frac{y^\top f(x)}{y^\top g(x)} \tag{4.25}$$

for every $x \in \mathcal{X}$ and $\Sigma := \{y \in \mathbb{R}^n : y \geq 0, \sum_{j \in J} y_j = 1\}$. This is a direct consequence of the property that a quasiconvex function attains its maximum in a vertex of a convex polyhedron, see [6]. Moreover, by the assumptions on the vector functions f and g we obtain that the function

$$x \longmapsto \frac{y^\top f(x)}{y^\top g(x)}$$

is quasiconvex on $\mathcal{X}$ for every $y \in \Sigma$, while the function

$$y \longmapsto \frac{y^\top f(x)}{y^\top g(x)}$$

is quasiconcave on Σ for every $x \in \mathcal{X}$. Hence, using the compactness of the convex sets $\mathcal{X}$ and Σ it follows from Sion's minimax theorem [128] that

$$\min_{x \in \mathcal{X}} \left\{ \max_{y \in \Sigma} \frac{y^\top f(x)}{y^\top g(x)} \right\} = \max_{y \in \Sigma} \left\{ \min_{x \in \mathcal{X}} \frac{y^\top f(x)}{y^\top g(x)} \right\}, \tag{4.26}$$

and so by (4.25) and (4.26) we obtain

$$\min_{x \in \mathcal{X}} \left\{ \max_{j \in J} \frac{f_j(x)}{g_j(x)} \right\} = \max_{y \in \Sigma} \left\{ \min_{x \in \mathcal{X}} \frac{y^\top f(x)}{y^\top g(x)} \right\}. \tag{4.27}$$

Let $c : \Sigma \longrightarrow \mathbb{R}$ be defined by

$$c(y) := \min_{x \in \mathcal{X}} \frac{y^\top f(x)}{y^\top g(x)}. \tag{4.28}$$

As shown in Lemma 3.1 of [108], the function c is continuous on Σ and moreover, it is semistrictly quasiconcave since it is the infimum of semistrictly quasiconcave functions, see [6]. Thus, by (4.27), we need to solve the quasiconcave optimization problem

$$\max_{y \in \Sigma} c(y) = \max_{y \in \Sigma} \left\{ \min_{x \in \mathcal{X}} \frac{y^\top f(x)}{y^\top g(x)} \right\} \tag{Q}$$

where a local maximum is a global maximum, see [6].

Problem (Q) can be seen as a "partial" dual program of the convex generalized fractional program (P), since it only "dualizes" the ratios. Moreover, contrary to the standard dual (D) we do not need to impose a Slater condition to ensure the solvability of (Q). In fact, it follows from (4.27) that there exists some $y_\star \in \Sigma$ with $c(y_\star) = \vartheta(Q) = \vartheta(P)$. However, this does not mean that the optimal solution $x^\star \in \mathcal{X}$ associated with the single-ratio fractional programming problem $\min_{x \in \mathcal{X}} \dfrac{y_\star^\top f(x)}{y_\star^\top g(x)}$ is also an optimal solution of (P). Notice also that optimization problem (Q) is a special generalized fractional program involving an infinite number of ratios.

It is now interesting to develop an algorithm capable of solving (Q). Due to the similarities between the standard dual (D) and the "partial" dual (Q), we will use the approach described in Section 4.2.1 to derive this new algorithm. Therefore, we will first relate (P_λ) to the parametric problem associated with (Q) program given by

$$\max_{y \in \Sigma} F(y, \lambda) \qquad\qquad (Q_\lambda)$$

with

$$F(y, \lambda) := \min_{x \in \mathcal{X}} \left\{ y^\top \left(f(x) - \lambda g(x) \right) \right\}.$$

Proposition 4.2.3

For $\lambda \in \mathbb{R}$ if $g : C \longrightarrow \mathbb{R}^n$ is an affine vector-valued function on $\mathcal{X}$ or for $\lambda \geq 0$ if $g : C \longrightarrow \mathbb{R}^n$ is a concave vector-valued function on $\mathcal{X}$, we have that

$$\max_{y \in \Sigma} F(y, \lambda) = F(\lambda) \qquad\qquad (4.29)$$

with F the value function associated with the parametric problem (P_λ) of (P).

Proof: By the assumptions on f and g and the compactness of the convex sets $\mathcal{X}, \Sigma$, we can apply Von Neumann's min-max theorem [115], and so

$$\max_{y \in \Sigma} F(y, \lambda) = \max_{y \in \Sigma} \left\{ \min_{x \in \mathcal{X}} \left\{ y^\top \left(f(x) - \lambda g(x) \right) \right\} \right\}$$

$$= \min_{x \in \mathcal{X}} \left\{ \max_{y \in \Sigma} \left\{ y^\top \left(f(x) - \lambda g(x) \right) \right\} \right\} = \min_{x \in \mathcal{X}} \left\{ \max_{j \in J} \left\{ f_j(x) - \lambda g_j(x) \right\} \right\}$$

$$= F(\lambda)$$

which concludes the proof. $\qquad\qquad\qquad\qquad\qquad\qquad\qquad\qquad\qquad\qquad\qquad \square$

Observe that if g is a concave vector-valued function, then by assumption $\vartheta(P)$ is nonnegative and therefore we are only interested in nonnegative values of the parameter λ. The above result shows that (Q_λ) corresponds to the Lagrangian dual of the usual parametric problem (P_λ), and hence, $F(\lambda_k) = \vartheta(P_{\lambda_k}) = \vartheta(Q_{\lambda_k})$. Therefore, if y_k is an optimal solution of (Q_{λ_k}) then the function $F_{y_k} : \mathbb{R} \longrightarrow \mathbb{R}$ given by

$$F_{\boldsymbol{y}}(\lambda) := \min_{\boldsymbol{x} \in \mathcal{X}} \left\{ \boldsymbol{y}^\top \left(\boldsymbol{f}(\boldsymbol{x}) - \lambda \boldsymbol{g}(\boldsymbol{x}) \right) \right\} \tag{4.30}$$

approximates the "primal" parametric function F at λ_k from below, and $F_{y_k}(\lambda_k) = F(\lambda_k)$. As a result, the root of the equation $F_{y_k}(\lambda) = 0$ given by $c(y_k)$ yields a lower bound on $\vartheta(Q)$. These observations lead to the construction of Algorithm 4.4 to solve (Q).

Step 0. Take $\boldsymbol{y}_0 \in \Sigma$, compute $c(\boldsymbol{y}_0) = \min_{\boldsymbol{x} \in \mathcal{X}} \dfrac{\boldsymbol{y}_0^\top \boldsymbol{f}(\boldsymbol{x})}{\boldsymbol{y}_0^\top \boldsymbol{g}(\boldsymbol{x})}$ and let $k := 1$;

Step 1. Determine $\boldsymbol{y}_k := \arg \max_{\boldsymbol{y} \in \Sigma} F(\boldsymbol{y}, c(\boldsymbol{y}_{k-1}))$;

Step 2. **If** $F(\boldsymbol{y}_k, c(\boldsymbol{y}_{k-1})) = 0$

        Then $\boldsymbol{y}_{k-1}$ is an optimal solution with value $c(\boldsymbol{y}_{k-1})$ and **Stop**.

        Else GoTo *Step 3*;

Step 3. Compute $c(\boldsymbol{y}_k)$, let $k := k + 1$, and **GoTo** *Step 1*.

Algorithm 4.4: Modified "dual" algorithm.

Alike Lemma 4.2.1, it can easily be shown that the sequence $\{c(\boldsymbol{y}_k)\}_{k \geq 0}$ constructed by the above algorithm is strictly increasing. Moreover, the approximation function $F_{\boldsymbol{y}}$ has the same type of properties as the function $G_{(\boldsymbol{y},\boldsymbol{z})}$, i.e. it is concave, continuous and decreasing, and this yields the geometrical interpretation of this modified "dual" algorithm in Figure 4.4.

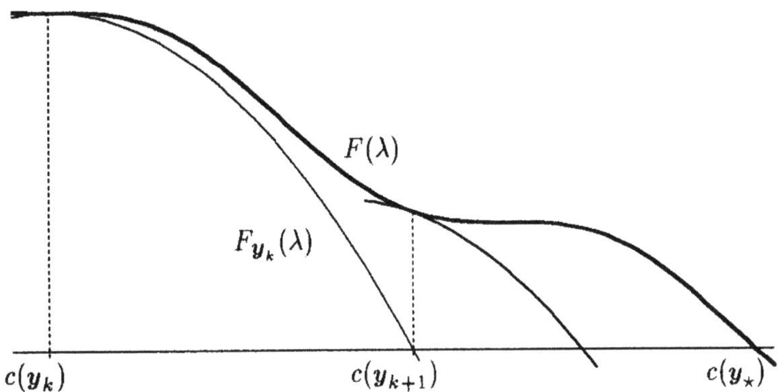

Figure 4.4: Geometric interpretation of the modified "dual" algorithm.

The convergence proof of this modified "dual" algorithm is similar to the one presented in the previous section and therefore it is omitted.

Theorem 4.2.2 ([14])

The sequence $y_k, 0 \leq k < k^\star$, does not contain optimal solutions of (Q) and the corresponding function values $c(y_k), 0 \leq k < k^\star$ are strictly increasing. Moreover, if $k^\star$ is finite, then $c(y_{k^\star}) = \lambda_\star = \vartheta(Q)$ while for $k^\star = +\infty$ we have

$$\lim_{k\uparrow\infty} c(y_k) = \lambda_\star.$$

Furthermore, the convergence rate of this algorithm is linear.

Contrary to the "dual" algorithm, we do not have to impose any type of Slater condition like in Lemma 4.2.3 to ensure the existence of accumulation points of the sequence $\{y_k\}_{k \geq 0}$. This is due to $y_k \in \Sigma$ with Σ a compact set. The main difference lies in the formats of the standard dual (D) and the modified "dual" problem (Q). The latter is a partial dual while the standard dual (D) is formed by not only dualizing the ratios but also some of the constraints of the feasible set $\mathcal{X}$. Therefore, in the standard dual it is required to ensure that the set of the optimal Lagrangian multipliers associated with these constraints is bounded. Observe that this is guaranteed by the strong Slater condition, see [73].

Lemma 4.2.6

The sequence $\{y_k\}_{k \geq 0}$ has an accumulation point $y_\star$ and this accumulation point is an optimal solution of (Q).

We can now derive a sufficient condition to achieve superlinear convergence of Algorithm 4.4.

Proposition 4.2.4 ([14])

If for every optimal solution $y_\star$ of (Q), the optimization problem

$$\min_{x \in \mathcal{X}} \frac{y_\star^{\mathsf{T}} f(x)}{y_\star^{\mathsf{T}} g(x)} \qquad (Q_\star)$$

has a unique optimal solution then the modified "dual" algorithm converges superlinearly.

The next corollary establishes sufficient conditions on the functions f_j and g_j to ensure that the convergence rate of the modified "dual" algorithm is superlinear. Contrary to the "dual" algorithm, see Corollary 4.2.1, these are less restrictive.

Corollary 4.2.2

If one of the following conditions holds

(i) $f : \mathcal{C} \longrightarrow \mathbb{R}^n$ is nonnegative and strictly convex on $\mathcal{X}$ and $g : \mathcal{C} \longrightarrow \mathbb{R}^n$ is positive and concave on $\mathcal{X}$;

(ii) $f : \mathcal{C} \longrightarrow \mathbb{R}^n$ is nonnegative and convex on $\mathcal{X}$ and $g : \mathcal{C} \longrightarrow \mathbb{R}^n$ is positive and strictly concave on $\mathcal{X}$;

(iii) $f : \mathcal{C} \longrightarrow \mathbb{R}^n$ is strictly convex on $\mathcal{X}$ and $g : \mathcal{C} \longrightarrow \mathbb{R}^n$ is positive and affine on $\mathcal{X}$

then the modified "dual" algorithm converges superlinearly.

In order to relate this duality approach with the one described in Section 4.2.1 we will start by analyzing their associated parametric problems. Observe that since the dual approach derived in the previous section requires that a Slater condition holds on $\mathcal{X}$, see Section 4.2.1, we will impose the same condition in order to relate the two approaches. Furthermore, we will also assume that the convexity assumptions of f and g also hold on $\mathcal{S}$.

Proposition 4.2.5

If the Slater condition holds and either $f : \mathcal{C}\longrightarrow \mathbb{R}^n$ is a convex vector-valued function on $\mathcal{S}$ and $g : \mathcal{C}\longrightarrow \mathbb{R}^n$ is a positive affine vector-valued function on $\mathcal{S}$ or $f : \mathcal{C}\longrightarrow \mathbb{R}^n$ is a convex vector-valued function on $\mathcal{S}$ and nonnegative on $\mathcal{X}$ and $g : \mathcal{C}\longrightarrow \mathbb{R}^n$ is a positive concave vector-valued function on $\mathcal{S}$. Then,

$$G(\lambda) = \max_{y \in \Sigma} F(y, \lambda) \tag{4.31}$$

for $\lambda \in \mathbb{R}$ if g is an affine vector-valued function or for $\lambda \geq 0$ if g is a concave vector-valued function. Moreover, $\widehat{y}$ is an optimal solution of (Q_λ) if and only if there exists some $\widehat{z} \geq 0$ such that $(\widehat{y}, \widehat{z})$ is an optimal solution of (D_λ).

Proof: Recall that by definition we have

$$G(\lambda) := \max_{y \in \Sigma} \max_{z \geq 0} \left\{ \min_{x \in \mathcal{S}} \left\{ y^\top f(x) + z^\top h(x) - \lambda y^\top g(x) \right\} \right\}.$$

Considering $z \geq 0$ as the vector of Lagrangian multipliers associated with the constraints $h(x) \leq 0$, we obtain by Theorem 28.2 of [117] that

$$\begin{aligned} F(y, \lambda) &= \min_{x \in \mathcal{X}} \left\{ y^\top f(x) - \lambda y^\top g(x) \right\} \\ &= \max_{z \geq 0} \min_{x \in \mathcal{S}} \left\{ y^\top \left(f(x) - \lambda g(x) \right) + z^\top h(x) \right\} \end{aligned} \tag{4.32}$$

and hence it follows that $G(\lambda) = \max_{y \in \Sigma} F(y, \lambda)$.

To verify the second part notice if $\widehat{y} \in \Sigma$ is an optimal solution of (Q_λ) that by (4.31) and (4.32) there exists some $\widehat{z} \geq 0$ satisfying

$$\begin{aligned} \max_{y \in \Sigma, z \geq 0} G(y, z, \lambda) = F(\widehat{y}, \lambda) &= \max_{z \geq 0} \left\{ \min_{x \in \mathcal{S}} \left\{ \widehat{y}^\top \left(f(x) - \lambda g(x) \right) + z^\top h(x) \right\} \right\} \\ &= G(\widehat{y}, \widehat{z}, \lambda) \end{aligned}$$

and so $(\widehat{y}, \widehat{z})$ is an optimal solution of (D_λ). Moreover, if $(\widehat{y}, \widehat{z})$ is an optimal solution of (D_λ) it follows by weak duality and (4.31) that

$$\max_{y \in \Sigma} F(y, \lambda) = G(\widehat{y}, \widehat{z}, \lambda) = \min_{x \in \mathcal{S}} \left\{ \widehat{y}^\top f(x) + \widehat{z}^\top h(x) - \lambda \widehat{y}^\top g(x) \right\} \leq F(\widehat{y}, \lambda)$$

and this shows that $\widehat{y}$ is an optimal solution of (Q_λ). $\qquad\qquad\square$

Notice from the above proposition that $\widehat{z}$ is the optimal Lagrangian multiplier vector associated with the constraints $h(x) \leq 0$ of the optimization problem defined by $F(\widehat{y}, \lambda)$.

In spite of the different formats of the duals (Q) and (D), they have by Proposition 4.2.5 equivalent associated parametric problems, if the Slater condition holds. Moreover, from Proposition 4.2.3 we also know that the parametric problems associated with (P) and (Q) are equivalent. Hence, we recover the equivalence relation between the parametric problems associated with (D) and (P) expressed in Proposition 4.2.1.

The above remarks permit also to show that *Step* 1 of the modified "dual" algorithm can be solved efficiently. From Proposition 4.2.5 it follows that y_{k+1} is an optimal solution of (Q_{λ_k}) if and only if (y_{k+1}, z_{k+1}) is an optimal solution of (D_{λ_k}). Therefore, assume that some constraint qualification constraint on $\mathcal{X}$ holds (like the Slater condition), the functions $f_j, g_j, j \in J$ are differentiable and that the convex functions defining the set $\mathcal{X}$ are also differentiable. In this case, solving *Step* 1 is equivalent to first solving (P_{λ_k}) and then finding a solution $\widehat{u}$ satisfying the associated Karush-Kuhn-Tucker system given by (4.18), (4.19), (4.20). In fact, from Lemma 4.2.4 we can derive an optimal solution of $(Q_{c(y_k)})$ using $\widehat{u}$.

From Proposition 4.2.5 it also follows that z_{k+1} belongs to

$$\arg \max \left\{ G(y_{k+1}, z, \lambda_k) : z \geq 0 \right\}.$$

This observation and the next result permit to rank the next iteration value λ_{k+1} of both methods.

Lemma 4.2.7

Assume that the Slater condition holds and that or $f : C \longrightarrow \mathbb{R}^n$ is a convex vector-valued function on S and $g : C \longrightarrow \mathbb{R}^n$ is a positive affine vector-valued function on S or that $f : C \longrightarrow \mathbb{R}^n$ is a convex vector-valued function on S and nonnegative on $\mathcal{X}$ and $g : C \longrightarrow \mathbb{R}^n$ is a positive concave vector-valued function on S. Then, for all $y \in \Sigma, z \geq 0$ the inequality $d(y, z) \leq c(y)$ is satisfied. Moreover, for (y_{k+1}, z_{k+1})

an optimal solution of (D_{λ_k}) *and* y_{k+1} *an optimal solution of* (Q_{λ_k}), $c(y_{k+1})$ *equals* $d(y_{k+1}, z_{k+1})$ *if and only if* z_{k+1} *belongs to*

$$\arg\max\left\{G(y_{k+1}, z, d(y_{k+1}, z_{k+1})) : z \geq 0\right\}.$$

Proof: By Lemma 3.1.1, we have

$$0 = \min_{x \in \mathcal{X}}\left\{y^{\top}\left(f(x) - c(y)g(x)\right)\right\}$$

and

$$0 = \min_{x \in \mathcal{S}}\left\{y^{\top}\left(f(x) - d(y, z)g(x)\right) + z^{\top}h(x)\right\}$$
$$\leq \min_{x \in \mathcal{X}}\left\{y^{\top}\left(f(x) - d(y, z)g(x)\right)\right\}.$$

Using again Lemma 3.1.1 the first result follows. For (y_{k+1}, z_{k+1}) an optimal solution of (D_{λ_k}) it follows by Lagrangian duality that the above inequality is actually an equality if and only if z_{k+1} belongs to $\arg\max\{G(y_{k+1}, z, d(y_{k+1}, z_{k+1})) : z \geq 0\}$. Hence, we obtain that $d(y_{k+1}, z_{k+1}) = c(y_{k+1})$ and the result is proved. $\qquad\square$

The above lemma raises the question if in practice the situation $d(y_{k+1}, z_{k+1}) < c(y_{k+1})$ occurs frequently. According to our computational experience this situation does occur at the beginning of the application of the algorithms. This is to be expected in view of Lemma 4.2.7.

It is left to discuss a stopping rule for the modified "dual" algorithm. However, due to Proposition 4.2.3, it follows that the stopping rule discussed in the previous section can also be applied for this algorithm.

Although the modified "dual" algorithm and Algorithm 4.3 are quite similar, it is important to analyze them in terms of the computational effort required at each iteration step. Observe that obtaining the next iteration point, i.e. solving $c(y_k)$ in the above algorithm, can be more difficult than in the case of Algorithm 4.3. In fact, whenever the feasible set $\mathcal{X}$ is formed by "easy" and "difficult" constraints, grouping in $\mathcal{S}$ the "easy" constraints can simplify the computation of the next iteration point in Algorithm 4.3, i.e. $d(y_k, z_k)$. On the other hand, the complexity of the solution procedure for *Step* 1 in both algorithms is identical, since both require the solution of the same parametric problem. Therefore, the apparent advantage of Algorithm 4.3 over the modified "dual" method whenever the feasible set $\mathcal{X}$ is formed by "easy" and "difficult" constraints, appears to be eliminated for this solution procedure. Another important issue is related to the sufficient conditions required by both algorithms to achieve superlinear convergence. In fact the sufficient conditions imposed by Corollary 4.2.1 for the "dual" algorithm are more restrictive.

Similar to the "dual" approach it is also possible to construct a scaled version the modified "dual" algorithm. The derivation of this scaled version is again rather technical, see [14], and produces the same type of convergence rate results. Moreover the scaled version does not improve the performance of its original version, see [14]. This may be explained, by the fact that both the "dual" algorithms are by themselves more robust and powerful than their primal counterpart.

We will now specialize the results derived to the linear case.

Linear case

The notation and assumptions used are similar to those of Section 4.2.1.

The new dual of (LP) is given by

$$\max_{y \in \Sigma} c(y) = \max_{y \in \Sigma} \left\{ \min_{x \in \mathcal{X}} \frac{y^\top (A^\top x + \alpha)}{y^\top (B^\top x + \beta)} \right\} \qquad (LQ)$$

and its associated parametric problem corresponds to

$$\max_{y \in \Sigma} \left\{ y^\top (\alpha - \lambda \beta) + \left\{ \min_{x \in \mathcal{X}} y^\top (A - \lambda B)^\top x \right\} \right\}. \qquad (LQ_\lambda)$$

Observe that using linear duality, the above problem can be reduced to the following linear programming problem

$$\max \ (\alpha - \lambda \beta)^\top y - \gamma^\top z$$

$$\text{s.t.: } (a_{i.} - \lambda b_{i.}) y + c_{i.} z \geq 0 \qquad i = 1, \ldots, m$$

$$y \in \Sigma, z \geq 0$$

which is the dual problem of the usual parametric problem associated with (LP). Hence, *Step* 1 of Algorithm 4.4 reduces to solving (LQ_λ).

Specializing Proposition 4.2.4 to the linear case boils down to demanding the uniqueness of the optimal solutions of the fractional programs

$$\min_{x \in \mathcal{X}} \frac{y_\star^\top (A^\top x + \alpha)}{y_\star^\top (B^\top x + \beta)} \qquad (LQ_\star)$$

defined for all optimal solutions $y_\star$ of (LQ). Observe that the uniqueness of optimal solutions of the above problems is equivalent to the uniqueness of the optimal solution of the associated parametric problem, with $\lambda_\star = \vartheta(LP)$

$$\min_{x \in \mathcal{X}} \left\{ y_\star^\top (A - \lambda_\star B)^\top x \right\}.$$

The above linear problem has a unique optimal solution if its dual problem

$$\max \left\{ -\gamma^\top z : -C^\top z \leq (A - \lambda_\star B)^\top y_\star, z \geq 0 \right\}$$

is such that all optimal basic solutions are nondegenerate. Observe that this is guaranteed whenever $(Q_{\lambda_\star})$ has a unique optimal solution which is a nondegenerate basic solution. This immediately implies that $(P_{\lambda_\star})$ has a unique optimal solution. Hence, a sufficient condition to achieve superlinear convergence of the modified "dual" algorithm in the linear case is given in the following corollary.

Corollary 4.2.3

If both (LP) and (LQ) have a unique optimal solution then the convergence rate of Algorithm 4.4 is superlinear.

Contrary to the nonlinear case it does not appear to be possible to establish sufficient conditions to verify a priori whether the above corollary holds. Nevertheless, the condition imposed by Corollary 4.2.3 appears to be "less restrictive" than the one established for Algorithm 4.3 in the linear case. Moreover, notice that the sufficient condition of Corollary 4.2.3 resembles one of the conditions imposed by Borde and Crouzeix [29] to ensure that the convergence rate of the Dinkelbach-type-2 algorithm is superlinear, see [44].

Finally, we will consider the dual problem proposed in [42, 43, 79]. In these papers the following dual of the problem (LP) is derived under the assumptions that the feasible set $\mathcal{X}$ is nonempty (weaker assumption than (A_1)) and (A_2)

$$\sup \left\{ t : t(\beta - \alpha)^\top y + \gamma^\top z \leq 0, t(B - A)y - Cz \leq 0, y \in \Sigma, z \geq 0 \right\}. \qquad (GLD)$$

According to Crouzeix *et al.* [43] an optimal solution of (GLD) can be found by solving the usual parametric problem (P_λ). However, this knowledge is not used to construct an algorithm to solve (GLD). On the other hand, as we just saw, by additionally assuming that the feasible set $\mathcal{X}$ is bounded we can apply the modified "dual" algorithm and consequently solve (LP) efficiently by means of the new dual problem (LQ).

4.2.3 Computational Results

In order to test the efficiency of these new "dual" methods introduced in Sections 4.2.1 and 4.2.2, we compared them with the Dinkelbach-type algorithm. The test problems considered are examples of convex generalized fractional programs and have in the numerator of the ratios quadratic functions, $f_j(x) := \frac{1}{2}x^\top H_j x + a_j^\top x + b_j$, and in the denominator linear functions, $g_j(x) := c_j^\top x + d_j$. The quadratic functions f_j are generated in the following way

- In the linear term each element of the vector a_j is uniformly drawn from $[-15.0, 45.0]$. Similarly, b_j is drawn from $[-30.0, 0]$;

- The Hessian is defined by $H_j := L_j U_j L_j^\top$ where L_j is a unit lower triangular matrix with components uniformly drawn from $[-2.5, 2.5]$ and U_j is a positive diagonal matrix with elements uniformly drawn from $[0.1, 1.6]$. Whenever a positive semidefinite Hessian is required the first component of the diagonal matrix is set to zero.

Finally, the linear functions g_j are constructed as follows: each element of the vector c_j is uniformly drawn from $[0.0, 30.0]$. Similarly, d_j is drawn from $[5.0, 35.0]$.

The feasible sets considered are the following:

$$\mathcal{X}_1 := \left\{ x \in I\!R^m : \sum_{i=1}^m x_i \le 1, 0 \le x_i \le 1, i = 1, \ldots m \right\}$$
$$\mathcal{X}_2 := \left\{ x \in I\!R^m : \sum_{i \in I_1} x_i \le 1, \sum_{i \in I_2} x_i \le 1, 0 \le x_i \le 1, i = 1, \ldots m \right\}$$

where $I_1 := \{1 \le i \le m : i \text{ is odd}\}$, $I_2 := \{1 \le i \le m : i \text{ is even}\}$. For both feasible sets, the "dual" algorithm considered the set $\mathcal{S}$ given by $\{x \in I\!R^m : 0 \le x_i \le 1, i = 1, \ldots, m\}$.

The three methods were implemented in Sun Pascal, linked to a pair of existing routines written in Sun FORTRAN and run on a Sun Sparc System 600 workstation using the default double precision (64-bit IEEE floating point format) real numbers of Sun Pascal and FORTRAN. Both compilers were used with the default compilation options.

For the minimization of the maximum of quadratic functions with linear constraints we used the bundle trust method coded in FORTRAN, see [101]. In both "dual" type algorithms *Step* 1, is solved by computing the correspondent minimal ellipsoidal norm problem, see Section 4.2.1. The fractional programming problem that occurs in both *Steps* 0 and 3 of the "dual" type algorithms is solved by Dinkelbach's algorithm, see Section 3.1. The code used to solve the above quadratic problems is an implementation in FORTRAN of Lemke's algorithm, see [112].

In the "dual" algorithm we used in *Step* 0 as initial points $y_0^\mathsf{T} := (\frac{1}{n}, \ldots, \frac{1}{n})$ and $z_0^\mathsf{T} := (0, \ldots, 0)$, while for the modified "dual" algorithm the initial point in *Step* 0 is given by $y_0^\mathsf{T} := (\frac{1}{n}, \ldots, \frac{1}{n})$. Finally, in the Dinkelbach-type algorithm we take in *Step* 0

$$\lambda_1 := c(y_0) = \min_{x \in \mathcal{X}} \frac{y_0^\mathsf{T} f(x)}{y_0^\mathsf{T} g(x)}.$$

The tolerance used in the stopping rule in all implementations is $\varepsilon := 5 \times 10^{-6}$. Due to the way the problems are generated, this implies that the accuracy of the solution values obtained is at least 10^{-6}.

The results of the computational experience are summarized in the following tables. For each pair (m, n), where m is the number of variables and n the number of ratios, 5 uncorrelated instances of the problem were generated and solved by the two algorithms. Hence, the entries of the tables are averages of the corresponding values. The columns under *Dinkel.* report the results obtained using the Dinkelbach-type algorithm, see Algorithm 4.1. Similarly, the columns under *"Dual"* report the results obtained using the "dual" algorithm, Algorithm 4.3, while *"MDual"* report the results obtained using the modified "dual" algorithm, Algorithm 4.4. In these cases two extra columns are presented concerning important steps of these algorithms. Hence, column *%Fr* refers to the percentage of the time used to compute the next iteration point, i.e. $d(y_k, z_k)$, respectively $c(y_k)$, while Column *%K* refers to the percentage of the time used to solve the Karush-Kuhn-Tucker system and thus obtaining (y_{k+1}, z_{k+1}), respectively y_{k+1}. The column *It* refers to the number of iterations performed by the corresponding algorithm. Each *Sec* column refers to the execution time in seconds of the Sun Sparc Station measured by the available standard **clock** function of the Sun Pascal library. This measures the elapsed execution time from the start to the end of the corresponding method, excluding input and output operations. In order to easily compare the three different methods we report for both feasible sets, under the columns *%Imp.* $\mathcal{X}_1$ and *%Imp.* $\mathcal{X}_2$, the percentages of improvement in total execution time between them. Thus, for each feasible set, the percentages of improvement in total execution time of the "dual" type algorithms over the Dinkelbach-type algorithm are contained in column *DD*, i.e. $(1 - \frac{Time(Dual)}{Time(Din)}) \times 100$ and in column *MDD*, i.e. $(1 - \frac{Time(MDual)}{Time(Din)}) \times 100$. Finally, column *Duals* contains the percentage of improvement in total execution time of the modified "dual" algorithm over the "dual" algorithm, i.e. $(1 - \frac{Time(MDual)}{Time(Dual)}) \times 100$.

Tables 4.1, 4.2 and 4.3 contain the results obtained for test problems where the quadratic functions f_j are strictly convex.

Prob.		*Dinkel.*		"*Dual*"				"*MDual*"			
m	*n*	*It*	*Sec*	*It*	*%Fr*	*%K*	*Sec*	*It*	*%Fr*	*%K*	*Sec*
5	5	7	0.49	5	14.1	3.6	0.77	3	13.7	0.0	0.50
10	5	9	5.81	5	10.1	1.4	4.65	3	10.3	1.2	3.21
15	5	9	13.32	5	16.6	2.4	10.88	2	19.4	1.5	7.40
20	5	8	37.91	6	8.3	0.9	43.01	3	9.6	0.6	21.53
5	10	8	0.88	5	11.8	4.6	0.83	3	16.0	2.9	0.55
10	10	13	10.96	6	8.5	1.5	7.02	3	8.6	0.8	4.62
15	10	11	22.27	6	13.2	2.0	15.90	3	11.0	1.3	9.59
20	10	9	43.01	6	8.2	0.8	44.86	3	9.5	0.7	26.81
5	15	7	1.61	6	14.1	3.4	1.11	3	9.5	4.4	0.59
10	15	12	12.95	7	8.6	1.3	9.84	3	9.7	1.0	5.12
15	15	10	25.06	6	9.4	1.1	20.90	3	11.0	1.1	11.99
20	15	12	77.72	7	8.0	0.8	66.90	3	8.7	0.6	32.42
5	20	7	1.04	7	8.9	4.7	1.13	3	11.6	3.2	0.67
10	20	11	12.78	7	10.2	1.9	7.84	4	11.8	1.8	5.14
15	20	12	38.04	7	8.0	1.0	25.68	3	9.4	0.8	14.75
20	20	13	88.22	7	8.2	0.8	65.50	3	9.7	0.7	34.68

Table 4.1: $\mathcal{X}_1$ and strictly quasiconvex ratios.

The fact that the functions f_j are strictly convex ensures, by Corollary 4.2.2, that the convergence rate of the modified "dual" algorithm is superlinear for these cases. Observe that, due to the characteristics of the feasible sets $\mathcal{X}_1$ and $\mathcal{X}_2$, the strong Slater condition is verified. Hence, also for these examples the convergence rate of the "dual" algorithm is superlinear, see Corollary 4.2.1.

Prob.		*Dinkel.*		*"Dual"*				*"MDual"*			
m	*n*	*It*	*Sec*	*It*	*%Fr*	*%K*	*Sec*	*It*	*%Fr*	*%K*	*Sec*
5	5	7	2.06	5	9.8	3.4	1.40	3	13.8	1.2	0.66
10	5	10	10.38	5	8.3	0.8	6.83	2	9.8	0.6	3.92
15	5	9	19.35	5	10.3	1.1	14.45	3	13.1	1.0	9.69
20	5	8	38.76	6	8.5	0.8	43.31	3	10.8	0.7	22.46
5	10	11	2.13	5	10.0	4.6	1.02	3	12.5	1.5	0.67
10	10	10	9.51	6	7.8	0.9	8.25	3	8.5	0.6	4.92
15	10	11	34.80	6	8.5	0.7	25.53	3	9.6	0.6	15.91
20	10	10	62.78	5	8.1	0.7	46.02	3	9.9	0.6	31.92
5	15	7	0.83	5	14.0	5.8	0.85	2	16.2	2.2	0.55
10	15	10	9.97	6	8.4	1.1	8.29	3	11.2	1.1	4.68
15	15	8	34.78	6	6.6	0.5	31.36	3	7.9	0.4	19.73
20	15	10	69.34	6	8.7	0.8	59.81	3	9.3	0.5	35.63
5	20	8	1.68	6	9.1	4.0	1.47	3	12.4	0.9	0.94
10	20	10	13.13	6	7.3	0.8	9.73	3	8.2	0.8	6.21
15	20	12	37.68	6	8.2	0.8	26.04	3	8.7	0.6	17.44
20	20	11	83.37	6	6.9	0.5	73.31	3	8.3	0.5	39.48

Table 4.2: $\mathcal{X}_2$ and strictly quasiconvex ratios.

From the results in Tables 4.1 and 4.2 it seems that the "dual" algorithm, Algorithm 4.3, is better in terms of number of iterations than the Dinkelbach-type algorithm. However, this improvement is not always as effective in terms of execution time, see Table 4.3. On average more iterations are required by the "dual" algorithm than the modified "dual" algorithm. Furthermore, the modified "dual" algorithm, Algorithm 4.4, has a much better performance than the "dual" algorithm, as Table 4.3 shows.

Prob.		% Imp. $\mathcal{X}_1$			% Imp. $\mathcal{X}_2$		
m	n	DD	MDD	Duals	DD	MDD	Duals
5	5	−56.6	−2.0	34.8	32.0	68.0	53.0
10	5	20.0	44.8	31.0	34.1	62.2	42.6
15	5	18.4	44.5	32.0	25.3	49.9	32.9
20	5	−13.4	43.2	49.9	−11.7	42.0	48.1
5	10	6.4	37.4	33.1	52.1	68.3	34.0
10	10	36.0	57.9	34.2	13.2	48.3	40.4
15	10	28.6	56.9	39.7	26.7	54.3	37.7
20	10	−4.3	37.7	40.2	26.7	49.2	30.6
5	15	31.0	63.2	46.7	−2.0	33.6	34.9
10	15	24.0	60.4	48.0	16.8	53.1	43.6
15	15	16.6	52.1	42.6	9.8	43.3	37.1
20	15	13.9	58.3	51.5	13.7	48.6	40.4
5	20	−9.3	35.7	41.2	12.5	43.9	35.9
10	20	38.7	59.8	34.4	25.9	52.7	36.1
15	20	32.5	61.2	42.6	30.9	53.7	33.0
20	20	25.8	60.7	47.0	12.1	52.6	46.1

Table 4.3: Summary for strictly quasiconvex ratios.

Tables 4.4, 4.5 and 4.6 contain the results obtained for test problems where the quadratic functions f_j are only convex. The results presented in these tables show that the behavior of the "dual" algorithm worsens in the case where the functions f_j are no longer strictly convex. In fact, both in terms of number of iterations and execution time, the performance of the "dual" algorithm is not so often better than the one of the Dinkelbach-type algorithm, especially for the feasible set $\mathcal{X}_1$. However, the modified "dual" algorithm has a better performance than both the "dual" algorithm and the Dinkelbach-type algorithm.

Another interesting observation is the fact that solving a simpler fractional program in Algorithm 4.3, i.e. computing $d(y_k, z_k)$, appears to take less time than computing $c(y_k)$ in Algorithm 4.4, see column %Fr. However, this improvement is not enough to compensate the usual bigger number of iterations performed by the "dual" algorithm. This phenomenon is more striking for the test problems with semistrictly quasiconvex ratios.

Prob.		Dinkel.		"Dual"				"MDual"			
m	n	It	Sec	It	%Fr	%K	Sec	It	%Fr	%K	Sec
5	5	9	1.75	10	9.9	3.1	2.37	3	8.0	0.9	0.83
10	5	12	10.56	6	11.0	2.3	4.81	3	11.8	1.4	2.39
15	5	11	20.84	10	8.6	1.8	25.37	3	8.5	1.4	9.13
20	5	10	37.70	6	9.6	1.1	33.02	3	9.6	0.9	18.39
5	10	10	1.86	11	8.8	3.7	2.47	3	12.7	2.2	0.77
10	10	8	7.01	10	9.3	1.8	8.77	3	8.0	1.4	3.65
15	10	9	30.80	7	13.6	2.8	22.25	3	12.5	2.8	9.09
20	10	11	46.47	8	11.3	1.6	44.20	3	12.2	1.7	21.58
5	15	7	3.24	9	4.4	2.1	3.88	4	4.6	1.2	2.12
10	15	12	14.79	8	7.9	1.2	10.21	3	6.8	1.2	4.53
15	15	12	44.00	9	6.8	1.1	32.34	3	5.8	0.9	14.92
20	15	11	70.57	8	7.9	1.0	58.95	3	7.6	0.8	29.80
5	20	11	1.18	11	10.4	3.1	1.76	4	14.4	2.3	0.67
10	20	11	14.09	9	8.0	1.5	11.86	3	9.7	1.5	4.99
15	20	14	38.62	7	9.5	1.4	23.54	3	7.4	1.3	11.55
20	20	11	82.36	9	7.4	0.9	77.81	3	6.2	0.7	34.61

Table 4.4: $\mathcal{X}_1$ and semistrictly quasiconvex ratios.

Prob.		Dinkel.		"Dual"				"MDual"			
m	n	It	Sec	It	%Fr	%K	Sec	It	%Fr	%K	Sec
5	5	9	1.88	7	19.6	5.4	1.84	3	22.5	1.1	1.01
10	5	12	11.87	7	8.8	1.7	6.70	2	11.3	1.2	3.03
15	5	9	23.77	11	6.8	0.9	31.70	3	9.5	0.8	11.17
20	5	10	39.98	5	9.2	0.8	35.82	3	10.4	0.6	23.35
5	10	10	1.44	9	10.3	4.0	1.76	4	13.5	3.1	0.96
10	10	8	10.12	7	9.8	1.3	8.95	3	10.2	0.6	4.35
15	10	10	24.71	7	8.9	1.2	23.37	3	9.7	1.1	11.44
20	10	10	53.55	6	8.0	0.7	53.19	3	8.9	0.5	31.85
5	15	9	4.52	6	11.4	3.6	1.19	3	12.6	1.6	0.98
10	15	10	10.51	8	9.4	1.4	9.78	3	11.5	1.1	4.89
15	15	11	38.45	7	6.3	0.6	35.31	3	7.0	0.5	19.23
20	15	9	59.60	7	7.4	0.8	66.00	3	8.8	0.6	32.70
5	20	11	4.51	9	6.5	1.1	2.97	4	8.4	2.5	1.43
10	20	10	13.98	9	7.3	1.4	14.70	3	8.6	1.0	5.66
15	20	10	36.00	8	6.9	0.8	34.94	3	6.7	0.6	18.51
20	20	11	87.38	11	6.5	0.6	99.90	3	6.9	0.4	42.30

Table 4.5: $\mathcal{X}_2$ and semistrictly quasiconvex ratios.

Prob.		% Imp. $\mathcal{X}_1$			% Imp. $\mathcal{X}_2$		
m	n	*DD*	*MDD*	*Duals*	*DD*	*MDD*	*Duals*
5	5	−35.7	52.5	65.0	2.2	46.0	44.8
10	5	54.5	77.4	50.3	43.6	74.5	54.8
15	5	−21.7	56.2	64.0	−33.4	53.0	64.8
20	5	12.4	51.2	44.3	10.4	41.6	34.8
5	10	−32.8	58.6	68.8	−22.7	33.0	45.4
10	10	−25.1	48.0	58.4	11.5	57.0	51.4
15	10	27.8	70.5	59.1	5.4	53.7	51.0
20	10	4.9	53.6	51.2	0.7	40.5	40.1
5	15	−19.5	34.6	45.3	73.7	78.3	17.7
10	15	31.0	69.4	55.6	6.9	53.4	50.0
15	15	26.5	66.1	53.9	8.2	50.0	45.5
20	15	16.5	57.8	49.5	−10.8	45.1	50.5
5	20	−49.6	42.8	61.8	34.3	68.3	51.8
10	20	15.8	64.6	57.9	−5.2	59.5	61.5
15	20	39.1	70.1	50.9	2.9	48.6	47.0
20	20	5.5	58.0	55.5	−14.3	51.6	57.7

Table 4.6: Summary for semistrictly quasiconvex ratios.

Although each iteration of the "dual" type algorithms is more "expensive" in terms of execution time than one of the Dinkelbach-type algorithm, this extra effort is compensated in the total time used, as Tables 4.3 and 4.6 show. Finally, it also appears that the number of variables has a decisive influence in the behavior of the Dinkelbach-type algorithm. This effect is also noticable for the "dual" type algorithms although in a smaller scale.

4.3 Conclusions

Lately, the emphasis on fractional programming has turned towards generalized fractional programming. This is mainly due to the large spectrum of applications which can be modeled in this way. The most well-known examples are given by economic equilibrium problems, management applications of goal programming and multi-objective programming involving ratios of functions. However, this set of examples can also be enlarged with an application in location analysis, see Section 4.1.2. Actually, we believe that there exist many more real life location models that belong to this special class of nonlinear programming problems. We hope that the contents of this chapter will help to bridge the gap between location analysis and generalized fractional programming.

One of the basic assumptions of generalized fractional programming is the positivity assumption which ensures that the denominators of the ratios of the objective function are positive in the domain. However, there are some applications for which this assumption has to be explicitly imposed. This "inclusion" may create difficulties when applying the popular primal parametric approach discussed in Section 4.1.1. As mentioned, the efficiency of such procedures depends mostly on the characteristics of the associated parametric problem. Therefore, in Section 4.1.3 we analyzed this class of problems and showed how the primal characteristics of the parametric approach can be used to solve them efficiently.

The usual solution techniques for generalized fractional programs are basically procedures designed to solve the primal problem. This is mainly due to the "awkward" form of the standard dual problem of a generalized fractional program. However, as shown in Section 4.2.1, it is possible to use the available duality results in order to develop a new algorithm which solves this "awkward" dual in an efficient way. Moreover, this "dual" algorithm extends to the nonlinear case the Dinkelbach-type algorithm applied to the standard dual of a generalized linear fractional program. Therefore, it can be seen as an extension of the Dinkelbach-type algorithm to the nonlinear case with a "difficult" parametric problem. However, under some reasonable assumptions, it is possible to solve this parametric problem efficiently in the nonlinear case. Moreover, due to information provided by the standard dual problem, better convergence rate results for the new algorithm than for the Dinkelbach-type algorithm applied to the primal problem can be derived. Finally, the approach developed in Section 4.2.1 also shows that the standard duality results for the special case of generalized linear fractional programs can easily be derived by specializing the duality results of the more general nonlinear case.

In this chapter we also proposed another dual problem for a more general class of convex generalized fractional programming, which includes convex generalized fractional programming. Similar to the standard dual, this new dual problem can be efficiently solved via a "dual" type algorithm. However, less restrictive assumptions are required not only to guarantee the solvability of this new dual problem but also to derive some convergence rate results. In fact, sufficient conditions ensuring superlinear convergence of the modified "dual" algorithm are easier to verify than the ones needed for the "dual" algorithm. Furthermore, the computational experiments performed for quadratic-linear ratios and linear constraints show that its performance improves both the one of the "dual" algorithm as well as its primal counterpart, the Dinkelbach-type algorithm. We believe that these duality results as well as the algorithmic tools here presented may help to solve convex generalized fractional programs more efficiently. Moreover, it also illustrates the use of duality results for algorithmic purposes.

Five

Summary and Remarks

The first part of this book was dedicated to the analysis of discrete location models, in particular to the 2-level extensions of the uncapacitated facility location problem. We presented a new model for an uncapacitated 2-level location problem. This new model explores the relations among the facilities to be located by considering simultaneously the fixed costs of opening facilities in the two levels and also the fixed costs of having facilities operating together. As a result, the model generalizes known models in the literature, like the uncapacitated facility location problem, the 2-echelon uncapacitated facility location problem and the 2-level uncapacitated facility location problem. For our general model we discussed three different formulations and derived lower and upper bounds which were tested in branch and bound schemes. The specialization of these results to the 2-echelon uncapacitated facility location problem and the 2-level uncapacitated facility location problem improves the known existing results for these problems. Moreover, it also suggests the importance of deriving valid inequalities for these 2-level problems. The computational experience indicates that the 2-level types of problems have distinct behaviors. In particular, it appears that the "easiest" 2-level problems is the 2-echelon uncapacitated facility location problem, while the "hardest" is the 2-level uncapacitated facility location problem.

The mentioned models only consider the traditional criterion of maximizing the total net profit. Therefore, we also considered location models where the profitability index is maximized. Solving these discrete location problems where a ratio of two functions has to be optimized, requires not only the use of classical integer programming techniques but also of fractional programming techniques. Therefore, we started by analyzing some variants of the uncapacitated facility location problem with a ratio as objective function. Using basic concepts and results of fractional programming it was possible to identify some 1-level fractional location problems which can be solved in polynomial time on the size of the problem. As expected, the 2-level fractional location problems revealed to be more "intricate" than their 1-level counterpart. In fact, only for the fractional 2-echelon uncapacitated facility location problem we were able to identify the cases for which solving this 2-level fractional location problem corresponds to decomposing it into several 1-level fractional location problems. We believe that these results will motivate further investigation in the field of fractional location problems.

In this book we also considered generalized fractional programs. Until recently, applications of generalized fractional programming did not include location analysis. However, such applications do exist as we have shown. Moreover, we believe that more real life location models can be well modeled using generalized fractional programming. Therefore, we hope to have uncovered another research direction in location analysis. This belief is also justified by the new algorithmic and duality results presented. In fact, we have shown that the standard duality results known for convex generalized fractional programming can actually be used to construct an efficient algorithm. This new algorithm solves the standard dual of a convex generalized fractional program and at the same time provides a solution of the original (primal) problem. Moreover, due to its characteristics it can be seen as the "dual" algorithm of the well-known Dinkelbach-type algorithm. We also proposed another dual problem for a more general class of convex generalized fractional programs. Similar to the standard dual, we also derived a "dual" type algorithm that efficiently solves this new dual. Among the advantages of this approach, is the fact that less restrictive assumptions are needed to ensure a superlinear convergence rate. Furthermore, the computational experiments performed for quadratic-linear ratios and linear constraints show that its performance improves both the one of the "dual" algorithm as well as its primal counterpart, the Dinkelbach-type algorithm.

Bibliography

[1] K. Aardal, M. Labbé, J. Leung, and M. Queyranne. On the 2-level uncapacitated facility location problem. *Informs Journal on Computing*, 8:289–301, 1996.

[2] R.K. Ahuja, T.L. Magnanti, and J.B. Orlin. *Network Flows. Theory, Algorithms, and Applications*. Prentice-Hall, Englewood Cliffs, New Jersey, 1993.

[3] Y.P. Aneja and S.N. Kabadi. Ratio combinatorial programs. *European Journal of Operational Research*, 81:629–633, 1995.

[4] K.M. Anstreicher. A monotonic projective algorithm for fractional linear programming. *Algorithimica*, 1:483–498, 1986.

[5] Y. Anzai. On integer fractional programming. *Journal of the Operations Research Society of Japan*, 17(1):49–66, 1974.

[6] M. Avriel, W.E. Diewert, S. Schaible, and I. Zang. *Generalized Concavity*, volume 36 of *Mathematical Concepts and Methods in Science and Engineering*. Plenum Press, New York, 1988.

[7] D.A. Babayev. Comments on the note of Frieze. *Mathematical Programming*, 7:249–252, 1974.

[8] M.L. Balinski and P. Wolfe. On Benders decomposition and a plant location problem. Technical Report ARO-27, Mathematica Inc., Princeton, New Jersey, 1963.

[9] A.I. Barros. *Discrete and Fractional Programming Techniques for Location Models*. PhD thesis, Tinbergen Institute Rotterdam, Amsterdam, 1995.

[10] A.I. Barros, R. Dekker, J.B.G. Frenk, and S. van Weeren. Optimizing a general optimal replacement model by fractional programming techniques. *Journal of Global Optimization*, 10:405–423, 1997.

[11] A.I. Barros, R. Dekker, and V. Scholten. A two-level network for recycling sand: A case study. Technical Report 9673/A, Econometric Institute, Erasmus University, 1996. To appear in European Journal of Operational Research.

[12] A.I. Barros and J.B.G. Frenk. Generalized fractional programming and cutting plane algorithms. *Journal of Optimization Theory and Applications*, 87(1):103–120, 1995.

[13] A.I. Barros, J.B.G. Frenk, and J. Gromicho. Fractional location problems. Technical Report TI 9-96-8, Tinbergen Institute Rotterdam, 1996. To appear in Location Science.

[14] A.I. Barros, J.B.G. Frenk, S. Schaible, and S. Zhang. A new algorithm for generalized fractional programs. *Mathematical Programming*, 72(2):147–175, 1996.

[15] A.I. Barros, J.B.G. Frenk, S. Schaible, and S. Zhang. Using duality to solve generalized fractional programming problems. *Journal of Global Optimization*, 8:139–170, 1996.

[16] A.I. Barros and M. Labbé. A note on the paper of Ro and Tcha. Working paper, 1992.

[17] A.I. Barros and M. Labbé. A general model for the uncapacitated facility and depot location problem. *Location Science*, 2(3):173–191, 1994.

[18] A.I. Barros and M. Labbé. The multi-level uncapacitated facility location problem is not submodular. *European Journal of Operational Research*, 72(3):607–609, 1994.

[19] R. Batta. Single server queueing-location models with rejection. *Transportation Science*, 22(3):209–216, 1988.

[20] J.E. Beasley. Lagrangean heuristics for location problems. *European Journal of Operational Research*, 65(3):383–399, 1993.

[21] C.R. Bector. Duality in nonlinear fractional programming. *Zeitschrift für Operations Research*, 17:183–193, 1973.

[22] O. Berman, R.C. Larson, and S.S. Chiu. Optimal server location on a network operating as an M/G/1 queue. *Operations Research*, 33(4):746–771, 1985.

[23] J.C. Bernard and J.A. Ferland. Convergence of interval-type algorithms for generalized fractional programming. *Mathematical Programming*, 43:349–364, 1989.

[24] O. Bilde and J. Krarup. Sharp lower bounds and efficient algorithms to the simple plant location. *Annals of Discrete Mathematics*, 1:79–87, 1977.

[25] G.R. Bitran and T.L. Magnanti. Duality and sensitivity analysis for fractional programs. *Operations Research*, 24(4):675–699, 1976.

[26] J. Bloemhof-Ruwaard. *Integration of Operations Research and Environmental Management*. PhD thesis, Wageningen Agricultoral University, P.O. Box 9101, 6700 HB Wageningen, the Netherlands, 1996.

[27] H.F. Bohnenblust, S. Karlin, and L.S. Shapley. Solutions of discrete two-person games. *Annals of Mathematics Studies*, 24:51–72, 1950.

[28] M. Boncompte and J.E. Martínez-Legaz. Fractional programming by lower subdifferentiability techniques. *Journal of Optimization Theory and Applications*, 68(1):95–116, 1991.

[29] J. Borde and J.P. Crouzeix. Convergence of a Dinkelbach-type algorithm in generalized fractional programming. *Zeitschrift für Operations Research*, 31:A31–A54, 1987.

[30] R.A. Brealey and S.C. Myers. *Principles of Corporate Finance*. McGraw-Hill, Singapure, 1988.

[31] A. Cambini and L. Martein. Equivalence in linear fractional programming. *Optimization*, 23:41–51, 1992.

[32] A. Cambini, S. Schaible, and C. Sodini. Parametric linear fractional programming for an unbounded feasible region. *Journal of Global Optimization*, 3:157–169, 1993.

[33] M.A.C. Cerveira. Generalized fractional programming and its application to a continuous location problem. Master's thesis, Econometric Institute, Erasmus University, P.O. Box 1738, 3000 DR Rotterdam, the Netherlands, 1993.

[34] A. Charnes and W.W. Cooper. Programming with linear fractional functionals. *Naval Research Logistics Quarterly*, 9:181–186, 1962.

[35] A. Charnes and W.W. Cooper. Goal programming and multiple objective optimizations. *European Journal of Operational Research*, 1:39–54, 1977.

[36] L. Cooper. Location-allocation problems. *Operations Research*, 11(3):331–343, 1963.

[37] L. Cooper. Heuristic methods for location-allocation problems. *SIAM Review*, 6(1):37–53, 1964.

[38] G. Cornuéjols, M.L. Fisher, and G.L. Nemhauser. Location of bank accounts to optimize float: an analytic study of exact and approximate algorithms. *Management Science*, 23:789–810, 1977.

[39] G. Cornuéjols, G.L. Nemhauser, and L.A. Wolsey. The uncapacitated facility location problem. In P.B. Mirchandani and R.L. Francis, editors, *Discrete Location Theory*, chapter 3. Wiley-Interscience, New York, 1990.

[40] B.D. Craven. *Fractional Programming*. Heldermann-Verlag, Berlin, 1988.

[41] J.P. Crouzeix and J.A. Ferland. Algorithms for generalized fractional programming. *Mathematical Programming*, 52(2):191–207, 1991.

[42] J.P. Crouzeix, J.A. Ferland, and S. Schaible. Duality in generalized linear fractional programming. *Mathematical Programming*, 27(3):342–354, 1983.

[43] J.P. Crouzeix, J.A. Ferland, and S. Schaible. An algorithm for generalized fractional programs. *Journal of Optimization Theory and Applications*, 47(1):35–49, 1985.

[44] J.P. Crouzeix, J.A. Ferland, and S. Schaible. A note on an algorithm for generalized fractional programs. *Journal of Optimization Theory and Applications*, 50(1):183–187, 1986.

[45] W. Dinkelbach. On nonlinear fractional programming. *Management Science*, 13(7):492–498, 1967.

[46] M.A. Effroymson and T.L. Ray. A branch-bound algorithm for plant location. *Operations Research*, 14(3):361–368, 1966.

[47] J. Elzinga, D. Hearn, and W. Randolph. Minimax multifacility location with euclidean distances. *Transportation Science*, 10(4):321–336, 1976.

[48] D. Erlenkotter. A dual-based procedure for uncapacitated facility location. *Operations Research*, 26(6):992–1009, 1978.

[49] J.A. Ferland and J.Y. Potvin. Generalized fractional programming: Algorithms and numerical experimentation. *European Journal of Operational Research*, 20:92–101, 1985.

[50] M.L. Fisher. The lagrangean relaxation method for solving integer programming problems. *Management Science*, 27(1):1–18, 1981.

[51] M.L. Fisher, G.L. Nemhauser, and L.A. Wolsey. An analysis on approximations for maximizing submodular set functions - II. *Mathematical Programming Study*, 8:73–87, 1978.

[52] J. Flachs. Generalized Cheney-Loeb-Dinkelbach-type algorithms. *Mathematics of Operations Research*, 10(4):674–687, 1985.

[53] B. Fox. Finding minimal cost-time ratio circuits. *Operations Research*, 17:546–551, 1969.

[54] R.W. Freund and F. Jarre. An interior-point method for fractional programs with convex constraints. *Mathematical Programming*, 67(3):407–440, 1994.

[55] R.W. Freund and F. Jarre. An interior-point method for multifractional programs with convex constraints. *Journal of Optimization Theory and Applications*, 85:125–161, 1995.

[56] R.W. Freund, F. Jarre, and S. Schaible. On self-concordant barrier functions for conic hulls and fractional programming. *Mathematical Programming*, 74(3):237–246, 1996.

[57] A.M. Frieze. A cost function property for plant location problems. *Mathematical Programming*, 7:245–248, 1974.

[58] S. Fujishige. *Submodular Functions and Optimization*, volume 47 of *Annals of Discrete Mathematics*. Elsevier Science Publisher B.V., Amsterdam, 1991.

[59] L.L. Gao and E.P. Robinson Jr. A dual-based optimization procedure for the two-echelon uncapacitated facility location problem. *Naval Research Logistics*, 39:191–212, 1992.

[60] L.L. Gao and E.P. Robinson Jr. Uncapacitated facility location: General solution procedure and computational experience. *European Journal of Operational Research*, 76(3):410–427, 1994.

[61] M.R. Garey and D.S. Johnson. *Computers and Intractability: A Guide to the Theory of NP-Completeness*. Freeman, San Francisco, 1979.

[62] A.M. Geoffrion. Lagrangian relaxation for integer programming. *Mathematical Programming Study*, 2:82–114, 1974.

[63] P.C. Gilmore and R.E. Gomory. A linear programming approach to the cutting stock problem-Part II. *Operations Research*, 11:863–888, 1963.

[64] D. Granot and F. Granot. On integer and mixed integer fractional programming problems. *Annals of Discrete Mathematics*, 1:221–231, 1977.

[65] M. Grunspan and M.E. Thomas. Hyperbolic integer programming. *Naval Research Logistics Quarterly*, 20:341–356, 1973.

[66] M. Gugat. A fast algorithm for a class of generalized fractional programs. *Management Science*, 42(10):1493–1499, 1996.

[67] P. Hansen, E.L. Pedrosa Filho, and C.C. Ribeiro. Location and sizing of offshore platforms for oil exploration. *European Journal of Operational Research*, 58(1):202–214, 1992.

[68] P. Hansen, E.L. Pedrosa Filho, and C.C. Ribeiro. Modelling location and sizing of offshore platforms. *European Journal of Operational Research*, 72(3):602–605, 1994.

[69] P. Hansen, M.V. Poggi de Aragão, and C.C. Ribeiro. Hyperbolic 0–1 programming and query optimization in information retrieval. *Mathematical Programming*, 52(2):255–263, 1991.

[70] S. Hashizume, M. Fukushima, N. Katoh, and T. Ibaraki. Approximation algorithms for combinatorial fractional programming problems. *Mathematical Programming*, 37:255–267, 1987.

[71] M. Held and R.M. Karp. The travelling-salesman problem and minimum spanning trees. *Operations Research*, 18:1138–1162, 1970.

[72] M.H. Held, P. Wolfe, and H.D. Crowder. Validation of subgradient optimization. *Mathematical Programming*, 6(1):62–88, 1974.

[73] J.B. Hiriart-Urruty and C. Lemaréchal. *Convex Analysis and Minimization Algorithms I: Fundamentals*, volume 1. Springer-Verlag, Berlin, 1993.

[74] J.B. Hiriart-Urruty and C. Lemaréchal. *Convex Analysis and Minimization Algorithms II: Advanced Theory and Bundle Methods*, volume 2. Springer-Verlag, Berlin, 1993.

[75] T. Ibaraki. Parametric approaches to fractional programs. *Mathematical Programming*, 26:345–362, 1983.

[76] J.R. Isbell and W.H. Marlow. Attrition games. *Naval Research Logistics Quarterly*, 3:71–93, 1956.

[77] H. Ishii, T. Ibaraki, and H. Mine. Fractional knapsack problems. *Mathematical Programming*, 13:255–271, 1977.

[78] R. Jagannathan. Duality for nonlinear fractional programs. *Zeitschrift für Operations Research*, 17:1–3, 1973.

[79] R. Jagannathan and S. Schaible. Duality in generalized fractional programming via Farkas' lemma. *Journal of Optimization Theory and Applications*, 41(3):417–424, 1983.

[80] A.V. Karzanov. *On minimal mean-cuts and circuits in a diagraph*, pages 72–83. Yaroslavl State Unviersity, 1985. In Russian.

[81] L. Kaufman, M.V. Eede, and P. Hansen. A plant and warehouse location problem. *Operations Research Quarterly*, 28:547–554, 1977.

[82] S. Kim and H. Ahn. Convergence of a generalized subgradient method for nondifferentiable convex optimization. *Mathematical Programming*, 50:75–80, 1991.

[83] K.C. Kiwiel. An aggregate subgradient method for nonsmooth convex optimization. *Mathematical Programming*, 27:320–341, 1983.

[84] J.G. Klincewicz, H. Luss, and E. Rousenberg. Optimal and heuristic algorithms for multiproduct uncapacitated facility location. *European Journal of Operational Research*, 26:251–258, 1986.

[85] J. Krarup and P.M. Pruzan. The simple plant location problem: Survey and synthesis. *European Journal of Operational Research*, 12:36–81, 1983.

[86] A.A. Kuehn and M.J. Hamburger. A heuristic program for locating warehouses. *Management Science*, 9:643–666, 1963.

[87] M. Labbé and R.E. Wendell. Localisation d'un dépôt de camions pour le transport de déchets. In *Gestion de l'Entreprise: l'Approche Quantitative*, pages 271–284. C.O.R.E., Brussels, 1988. In French.

[88] C. Lemaréchal. An extension of Davidon methods to nondifferentiable problems. *Mathematical Programming Study*, 3:95–100, 1975.

[89] A.S. Manne. Plant location under economies-of-scale decentralization and computation. *Management Science*, 11:213–235, 1964.

[90] B. Martos. Hyperbolic programming. *Naval Research Logistics Quarterly*, 11:135–155, 1964. English translation from: Publications of Mathematic Institute Hungarian Academy Sciences, 5:383-406, 1960.

[91] B. Martos. *Nonlinear Programming. Theory and Methods*. North-Holland, Amsterdam, 1975.

[92] N. Meggido. Combinatorial optimization with rational objective functions. *Mathematics of Operations Research*, 4(4):414–424, 1979.

[93] P.B. Mirchandani and R.L. Francis, editors. *Discrete Location Theory*. Discrete mathematics and optimization. Wiley-Interscience, New York, 1990.

[94] B. Mond. On algorithmic equivalence in linear fractional programming. *Mathematics of Computation*, 37(155):185–187, 1981.

[95] Y. Myung and D. Tcha. Return of investment analysis for facility location. Technical Report OR 251-91, Massachusetts Institute of Technology, 1991.

[96] G.L. Nemhauser and L.A. Wolsey. *Integer and Combinatorial Optimization*. Wiley-Interscience, 1988.

[97] G.L. Nemhauser, L.A. Wolsey, and M.L. Fisher. An analysis on approximations for maximizing submodular set functions - I. *Mathematical Programming*, 14:265–294, 1978.

[98] A.S. Nemirovsky. On polynomiality of the method of analytical centers for fractional programming. *Mathematical Programming*, 73:175–198, 1996.

[99] A.S. Nemirovsky. The longstep method of analytic centers for fractional problems. *Mathematical Programming*, 77:191–224, 1997.

[100] J.V. Neumann. A model of general economic equilibrium. *Review Economic Studies*, 13:1–9, 1945.

[101] J. Outrata, H. Schramm, and J.Zowe. Bundle trust methods: Fortran codes for nondifferentiable optimization. User's guide. Technical Report 269, Mathematisches Institut. Universität Bayreuth, 1991.

[102] C.H. Papadimitriou and K. Steiglitz. *Combinatorial Optimization, Algorithms and Complexity*. Prentice-Hall, 1982.

[103] P.M. Pardalos and A.T. Phillips. Global optimization of fractional programs. *Journal of Global Optimization*, 1(2):173–182, 1991.

[104] R.G. Parker and R.L. Rardin. *Discrete Optimization*. Academic Press, Boston, 1988.

[105] S.J. Phillips and M.I. Dessouky. The cut search algorithm with arc capacities and lower bounds. *Management Science*, 25(4):396–404, 1979.

[106] F. Plastria. Lower subdifferentiable functions and their minimization by cutting planes. *Journal of Optimization Theory and Applications*, 46(1):37–53, 1985.

[107] B.T. Poljak. A general method of solving extremum problems. *Soviet Mathematics Doklady*, 8:593–597, 1967.

[108] B.N. Pshenichnyi. *Necessary Conditions for an Extremum*. Marcel Dekker Inc., New York, 1971.

[109] T. Radzik. *Algorithms for some linear and fractional combinatorial optimization*. PhD thesis, Department of Computer Science, Standford Unviersity, Standford, U.S.A, 1992.

[110] T. Radzik. Newton's method for fractional combinatorial optimization. In *33rd IEEE Annual Symposium on Foundations of Computer Science*, pages 659–669, 1992.

[111] T. Radzik. Parametric flows, weighted means of cuts and fractional combinatorial optimization. In P.M. Pardalos, editor, *Complexity in Numerical Optimization*, pages 351–386. World Scientific Publishing, 1993.

[112] A. Ravindran. A computer routine for quadratic and linear programming problems. *Communications of the ACM*, 15(9):818, 1972.

[113] C. Revelle and G. Laporte. New directions in plant location. *Studies in Locational Analysis*, 5:31–58, 1993.

[114] H. Ro and D. Tcha. A branch-and-bound algorithm for the two-level uncapacitated facility location problem with some side constraints. *European Journal of Operational Research*, 18:349–358, 1984.

[115] A.W. Roberts and D.E. Varberg. *Convex Functions*. Academic Press, New York, 1973.

[116] P. Robillard. (0, 1) hyperbolic programming problems. *Naval Research Logistics Quarterly*, 18:47–57, 1971.

[117] R.T. Rockafellar. *Convex Analysis*. Princeton University Press, Princeton, New Jersey, 1970.

[118] R.T. Rockafellar. Generalized subgradients in mathematical programming. In A. Bachem, M. Grötschel, and B. Korte, editors, *Mathematical Programming. The State of the Art*, chapter 2, pages 368–390. Springer-Verlag, 1983.

[119] K.E. Rosing. Considering offshore production platforms. *European Journal of Operational Research*, 72(1):204–206, 1994.

[120] S. Schaible. Duality in fractional programming: A unified approach. *Operations Research*, 24(3):452–461, 1976.

[121] S. Schaible. Fractional programming. I, Duality. *Management Science*, 22:858–867, 1976.

[122] S. Schaible. Fractional programming. II, On Dinkelbach's algorithm. *Management Science*, 22:868–873, 1976.

[123] S. Schaible. Fractional programming: Applications and algorithms. *European Journal of Operational Research*, 7:111–120, 1981.

[124] S. Schaible. Fractional programming. In R. Horst and P.M. Pardalos, editors, *Handbook of Global Optimization*, volume 2 of *Nonconvex Optimization and Its Applications*, pages 495–608. Kluwer Academic Publishers, Dordrecht, 1995.

[125] S. Schaible and T. Ibaraki. Fractional programming (Invited Review). *European Journal of Operational Research*, 12:325–338, 1983.

[126] N.Z. Shor. *Method for minimizing nondifferentiable functions and their applications*. PhD thesis, Kiev, 1970.

[127] N.Z. Shor. *Minimization Methods for Non-differentiable Functions*, volume 3 of *Computational Mathematics*. Springer-Verlag, Berlin, 1985.

[128] M. Sion. On general minimax theorems. *Pacific Journal of Mathematics*, 8:171–176, 1958.

[129] M. Sniedovich. Fractional programming revisited. *European Journal of Operational Research*, 33:334–341, 1988.

[130] J.F. Stollsteimer. A working model for plant numbers and locations. *Journal of Farm Economics*, 45:631–645, 1963.

[131] D. Tcha and B. Lee. A branch-and-bound algorithm for the multi-level uncapacitated facility location problem. *European Journal of Operational Research*, 18:35–43, 1984.

[132] H.M Wagner and J.S.C. Yuan. Algorithmic equivalence in linear fractional programming. *Management Science*, 14(5):301–306, 1968.

[133] J. Werner. Duality in generalized fractional programming. In K.H. Hoffman, J.B. Hiriart-Urruty, C. Lemaréchal, and J. Zowe, editors, *Trends in Mathematical Optimization*, International Series of Numerical Mathematics, pages 197–232. Birkhäuser-Verlag Basel, 1988.

[134] P. Wolfe. A method of conjugate subgradients for minimizing nondifferentiable functions. *Mathematical Programming Study*, 3:145–173, 1975.

[135] A.S. Nemirovsky Yu.E. Nesterov. An interior-point method for generalized linear-fractional programming. *Mathematical Programming*, 69:177–204, 1995.

Index

Combinatorial Optimization

1. E. Çela: *The Quadratic Assignment Problem. Theory and Algorithms.* 1998
   ISBN 0-7923-4878-8
2. M.Sh. Levin: *Combinatorial Engineering of Decomposable Systems.* 1998
   ISBN 0-7923-4950-4
3. A.I. Barros: *Discrete and Fractional Programming Techniques for Location Models.*
   1998   ISBN 0-7923-5002-2

KLUWER ACADEMIC PUBLISHERS – DORDRECHT / BOSTON / LONDON

MIX
Papier aus verantwortungsvollen Quellen
Paper from responsible sources
FSC® C105338

If you have any concerns about our products,
you can contact us on
ProductSafety@springernature.com

In case Publisher is established outside the EU,
the EU authorized representative is:
**Springer Nature Customer Service Center GmbH
Europaplatz 3, 69115 Heidelberg, Germany**

Printed by Libri Plureos GmbH
in Hamburg, Germany